DOUBLE KNOT

DOUBLE KNOT

A WAR MEMOIR IN SEVEN ESSAYS

MAC CALTRIDER

Copyright © 2024 Mac Caltrider
First Edition | First Printing

All rights reserved.
This book or any portion thereof
May not be used or reproduced in any fashion whatsoever
Without the permission of the publisher or the author
Except for the use of quotations

Publisher: Dead Reckoning Collective
Book Cover Artwork & Design: Tyler James Carroll
Editor: Tyler James Carroll

Printed in the United States of America

Library of Congress Control Number: 2024935803

ISBN-13: 979-8-9862724-9-8 (paperback)

*This book is dedicated to the Gold Star families whose loved ones
died fighting America's longest war*

Double Knot	29
Dead Weight	47
Harden Up	61
When Pigs Fly	69
The Dog and Pony Show	91
The Mountain	119
Going Cigaretting	151

Foreword

I FIRST HEARD of Mac Caltrider in the spring of 2021 when my boss at *Coffee or Die Magazine* hired Mac as a staff writer. I was a senior editor for the outlet, and I assumed Mac was another of the many wet-behind-the-ears writers that gravitated to the publication like stray cats in need of a home.

The strays came in many stripes: first-time byline chasers, hardened warriors writing through their experiences, boot reporters hoping to learn the ropes, and the rarest of all breeds — the real-deal writers.

For all the just-passing-throughs whose work came across my desk at our halfway house for Hemingway-worshiping wannabes and O'Brien acolytes, there were a small few that fell into that final category of *real-deal writer*.

My preliminary investigation of Mac's bonafides revealed that he checked a lot of boxes. He was undoubtedly well read. The carefully curated Instagram page *Pipes & Pages* is a product of Mac's voracious literary appetite and an expression of his creative connection to words and images. I was intrigued to learn its founder — who seemed to subsist on a steady diet of great books and pipe tobacco — was joining my staff.

Mac had penned a few pieces for the magazine before he came on board as a staffer, but the first piece of his work I read was an unpublished draft of *When Pigs Fly*, the fourth essay in this memoir. Reading it, I suspected right away that the Universe was gifting me a wet ball of true talent that I could help mold into a Frankenstein monster of reportorial prowess and narrative depth. *All my dreams.*

Writers in their literary larval phase give aging editors something to get out of bed for in the morning. They're the ones who make the obsessive toil of journalistic storycraft worthwhile.

Mac Caltrider was the real deal.

What he lacked in formal training and experience, Mac made up for with a tireless work ethic and obsessive fidelity to writing and getting better every day. Over the period that he and I worked together, I saw his skill grow beyond my wildest dreams. It was my great honor as a storyteller to influence Mac's trajectory as a writer, and it remains my great honor to call him my friend.

Reserved and humble, Mac has a certain mystique. He isn't exactly what one might expect from a Marine combat veteran turned Baltimore cop.

There's a general rule when dealing with veterans and the war stories they tell (or don't tell): The boastful ones are mostly full of shit. The quiet ones — they've seen shit.

I always figured Mac had seen some shit, but his lens as a storyteller pointed mostly outward during his tenure as a journalist for our news and lifestyle magazine. I always hoped he'd arrive at the point of meaning-making within his own story and humbly deliver some hard-won literary wisdom to the world. With *Double Knot,* he has done exactly that. It's a piece of work that captures the holiest of literary grails — that thing we talk about when we talk about "higher Truth."

When I received my advanced copy of *Double Knot*, I tore through it and was not at all surprised to find that the book is easily one of the strongest pieces of creative nonfiction produced by a veteran of America's war on terror. Each of these seven essays delivers intellectual and emotional depth with structural and syntactic precision. They deftly convey war's infinite contradictions — the courage and cowardice, resilience and frailty, love and hatred, creation and destruction.

There are passages in this book that absolutely level me — moments where I feel the devastating depth of the suffering and loss from 20 years of war. But those moments are juxtaposed throughout with virtue, honor, humor and sacrifice. It's the kind of book that, as a writer, you sort of love and hate at the same time. You love it because you wish you'd written it. And you hate it because you wish you'd written it.

With this book, Mac has done the difficult, spiritual work of interrogating the narrative arc of his own hero's journey. As John Steinbeck so eloquently put it in *East of Eden*, "We have only one story. All novels, all poetry, are built on the neverending contest in ourselves of good and evil."

In *Double Knot*, that universal story shines like white phosphorus piercing the night sky above the graveyard of empires. What follows in these pages is an immaculate canvas rendered by the lamplight of a warrior's soul.

God bless Mac Caltrider for bearing witness and telling *the* story.

ETHAN ROCKE

BESTSELLING AUTHOR, EDITOR, AND WRITING COACH

FEBRUARY 21ST, 2024

Author's Note

SOME OF THESE essays first appeared in *Coffee or Die Magazine,* though the versions in this book are vastly different and communicate a different message.

Introduction

PEOPLE SOMETIMES SAY the war in Afghanistan didn't have a front line. But, if you knew where to look, the front line in Afghanistan was as obvious as a trench on the Western Front. Only, instead of barbed wire and dugouts, it was marked by remote outposts and squad-sized patrol bases. And, as in all wars, the front line in Afghanistan was established, expanded, and paid for by the infantry.

I was sitting in Mrs. Skylar's seventh-grade English class at the Boys' Latin School of Maryland when the price paid by the infantry first hit home. It was 2004, the war in Iraq was devolving into a bloodbath, and I was living in a bubble of ignorance half a world away from the violence. A teary-eyed teacher stuck her head in the classroom and asked Mrs. Skylar to step into the hall for a private word. When she returned, she was crying too. She told us that one of her former students, Nicholas "Ski" Ziolkowski, was killed while fighting in Iraq. His death — and more so the emotional bomb it set off within the elite all-boys college preparatory school — drove me to learn more about who Ski was, and how he wound up dying in a place none of us could find on a map.

The more I learned about Ski, the more I admired him. He'd been a well-liked cross country runner in high school. After graduation, he deviated from the traditional path of his classmates and decided to forgo college. While they attended some of the best universities in the country, he enlisted in the Marine Corps. He joined the infantry and became a Scout Sniper. Three years later, he was shot in the head and killed during the Second Battle of Fallujah. While I wasn't in a hurry to die, there was something undeniably attractive about leading a life so crammed with passion and selfless service that, in spite of being cut tragically short, it still united a place as small and often disconnected as a school.

Ski graduated from Boys' Latin the year before I got there, yet his sacrifice was sending waves through the school. What was it about Ski that had his former teachers openly crying? I came to learn it was his kindness, integrity, and determination — his character. Ski had every opportunity in the world, yet he felt called to serve. He could have pursued higher education or even become an officer. He qualified for every job in the military but chose to be a grunt. He chose to be among the most vital (and neglected) members of the military. That was what I wanted, too. Even then, I felt it in my bones.

I spent my senior year politely enduring a parade of family and friends who urged me to attend college instead of enlist. They all sang a similar tune. "You have too many opportunities," they'd say. "You're too smart. Too kind. Too creative, to throw your life away in the Marines." When that didn't work, they adjusted their strategy. "Why not be something special, like a Navy SEAL? Why not be an officer?" All questions I imagine Ski likely fielded in 2001.

They didn't understand that I wanted to be tested — not by a particularly hard brand of initial training, but by living the lowly existence of a grunt. I wanted to fight, and equally so, I wanted to live the Spartan lifestyle of the infantryman.

What I suspected was that the Marine Corps was filled with smart, kind, and creative young men. Men who also yearned to participate in the ultimate test to confirm what they'd always suspected about themselves. In the infantry, they wanted to experience that test at bayonet range. I craved the dirty, miserable life that only the grunt knows. I sought belonging, hardship, and adventure. The Marine Corps did not disappoint.

If it is true that we tell ourselves stories in order to live, that we look for the sermon in the suicide and the moral lesson in mass murder, then the immense amount of books that come out of every war should be of no surprise. If seeking a "narrative line" in life's entropy is the reason for writing about any topic, then what better subject to confront with a keyboard than something as broad and absurd as war?

After all, when all other forms of communication fail, societies send their children to kill strangers. If that weren't absurd enough, in Afghanistan, most of us kids wanted to kill said strangers and volunteered to do so. In fact, we worked hard and risked much to arrive on that contour of the map where the front line existed and most of the killing took place.

Like many veterans who pick up the pen after putting down the rifle, I felt a deep compulsion to write about the absurdities I'd participated in. Initially, I simply scribbled down significant moments the way I remembered them. I didn't put any thought into structure, word choice, or even who I was writing for. It was purely an act of recordkeeping. Writing eased my fear that if those moments didn't get preserved on paper, they would fade and ultimately evaporate.

Eventually, the stories I wrote began to take the form of a memoir. When it was done, I printed it out and buried it in my desk drawer alongside paperclips, dried-out highlighters, and the other detritus found in every writer's junk drawer. During the years it sat collecting dust, I was lucky enough to write for the now-defunct *Coffee or Die Magazine*. I worked alongside talented writers and editors during the magazine's meteoric rise from what began as little more than a blog to a digital publication that boasted millions of monthly subscribers and sold thousands of print magazines in the print edition's first year. The job not only paid my bills, it also gave me the tools and experience I needed to turn those stories I'd penned into something worth sharing. When I eventually returned to the memoir, I could see glimmers of good writing buried beneath 97,712 words of junk. After a few more years of separation from the war, and, more importantly, with a few years of writing professionally under my belt, I realized that what I wanted to say was best distilled down into essays. Additionally, they were better when cushioned by stories indirectly related to combat. By trimming the fat, I've tried to avoid writing a book bloated with glorified skirmishes and the banalities of military life.

These essays are not all overtly about my brief time at war, but my experiences in Afghanistan are the narrative line that connect them. My hope is that these seven stories provide a glimpse into what drives young people from different walks of life to voluntarily join an organization that doesn't "promise them a rose garden." Furthermore, I hope they convey what combat in Helmand Province at the height of the Afghanistan War was like for the average groundpounder. And finally, how that experience shapes the second act of their lives.

"They talk about the crack, the buzz, and the fix. They talk about having to have it, of being unable to forget it when they do, of not wanting to forget it — ever. They talk about being sustained by it, telling and retelling what happened and what it felt like. They talk about it with the pride of the privileged, of those who have had, seen, felt, been through something other people have not. They talk about it in the way that another generation talked about drugs or drink or both, except that they also use both drugs and drink. One lad, a publican, talks about it as though it were a chemical thing or a hormonal spray or some kind of intoxicating gas — once it's in the air, once an act of violence has been committed, other acts will follow inevitably — necessarily."

Among the Thugs, Bill Buford

Double Knot

> *"More than their willingness to fight was the reality of their continual injection into the most violent of places. That is the truest honor of the infantry; the casual, expendable nature of their existence."*
>
> No Joy, David Rose

CLIMBING UP THE 10-foot ladder to Post Four was always more inconvenient than I remembered. After I threw a case of water up to the sandbag and plywood room, whoever I was replacing would grab my rifle and offer a helping hand up the last rung. Ascending the rickety wood ladder wearing full kit took enough effort to bump up my heart rate in the sweltering late-April Afghan air.

The post was one of four. Each marked a corner of the rectangular combat outpost. The four HESCO walls of dirt-filled burlap and chicken wire cubes could withstand small arms, mortar, or rocket attacks. They also cost next to nothing, and since they didn't hurt the DOD's wallet, the American War Machine deemed HESCO the official building block of the GWOT.

"Five minutes late and I would have actually lost my mind," Broome said, bending to help me up the ladder's last rung. "I talked to myself for the last hour."

"Come on, babe, you know I'd never be late replacing you," I said. "You should've asked for relief over the radio; I could have brought you water. Or you could just start smoking like everyone else."

Broome wasn't smiling. He wasn't upset, but his wide eyes conveyed a level of exhaustion that rivaled that of a trucker at the end of a 500-mile haul. He gathered up two water bottles filled with amber liquid and stumbled past the burlap sheet that acted as the post's door.

"Keep an eye out for Jerry," Broome said as he climbed down. "Other than him, everything is kosher."

I opened my mouth to ask who the hell Jerry was, but decided to let Broome sleep. Clearly, he was hallucinating from six hours in the hot box. I could still hear him talking to himself when he wandered from the piss tubes back to the tent. *Poor guy still thinks he is talking to Jerry.*

With Broome gone, I set my rifle in the corner, kicked the waters to the side, and checked the M240 machine gun. I adjusted the five sandbags in the middle of the little cell to make them comfortable enough to sit on, then leaned into the gun's stock, peeped through the scope, and admired the view. Post Four offered a commanding view of the north and west. Directly in front was a gravel landing zone, bordered by a lone dirt road that stretched south to infinity. Parallel with it was a canal where children sometimes fetched water or washed their faces. Beyond that was an endless sea of poppies. At the foot of the post was a tangled line of concertina wire meant to prevent anyone from crawling close enough to toss a hand grenade inside. Once settled, I lit cigarette number one.

The sun was down, but it was still light enough to smoke without having to worry about snipers. Chances of a Taliban marksman scoring a headshot through the opening in the post's sandbags were small, but enough shots had been fired at the post for me to duck my head between my knees whenever I took a nighttime drag.

Sitting alone in a six-foot by six-foot dirt square in my gear for hours on end was enough to make me question my sanity. I could only smoke so many cigarettes or masturbate so many times each shift before I was forced to accept my place in Pashtun purgatory and suffer in silence. It was also the only place I could truly be alone, free to fantasize about whatever I wanted and let my mind wander in any direction it liked. On patrol, it took discipline to keep my mind from wandering. One lapse into daydreamy bliss and I might miss the only sign an IED was buried in front of me. On post, all I had to worry about was not falling asleep. I dreamt of getting back to the world, eating some pizza, and drinking beer. But mostly I thought about getting laid.

I made sure I had a full pack of cigarettes this time. With 20 smokes, I could burn three every hour and still have two to spare. I still wasn't used to standing post for six hours instead of the normal four. When we first arrived at the outpost, we only stood post for two hours at a time. It seemed long back then, but after Rory lost his leg we had to stretch it to four hours. Now that Cavalier was gone, too, we each stood six.

By the end of deployment, we had a designated guard force made up of shitbags from the battalion whom no one wanted to bring on patrol anymore. We all hated the guard force guys. The fact none of them came from our platoon made it easier to hate them. They didn't carry their own weight. In hindsight, I think a lot of those guys probably were not bad people — just bad Marines. But in war, bad Marines get people killed. Most of them probably suffered some kind of psychological break, and having them take over standing post freed the rest of us to focus on fighting and surviving. Full nights of sleep made both of those easier.

Six hours goes by painfully slowly when there's nothing to do but stare out across the edge of the empire. I tried not to admit it to myself, but part of me was pissed at Cavalier for getting blown up. As soon as the thought bumped against the periphery of my consciousness, I felt shame, and I swore I'd stand post for 12, 18, 24 fucking hours if it meant Cavalier was still with us.

He wasn't the first Marine to hit an IED. Our battalion had been in Afghanistan only three months, and finding IEDs was already becoming a common occurrence. Our squad had hit three. One took Rory's leg two weeks after we arrived. A few weeks later, a second IED destroyed West's foot. Cavalier hit the third a month later, but his was different. The IED he stepped on was a world-shaker.

The inside of a modern hand grenade, which is designed to kill anyone within 16 feet and wound anyone within 50, holds about six ounces of explosives. The IED Cavalier stepped on held 25 *pounds*. That much power should have made the entire squad disappear.

I was standing about 20 yards behind him when it detonated. His left leg disintegrated from the hip down. His right leg was severed halfway down the femur. The only thing left of it was his foot, ankle, and a few inches of calf. That part remained neatly secured in its boot, laces still tied in a neat double knot. It landed in a cornfield 100 yards away.

The blast also collapsed one of Cavalier's lungs, ripped his right ass cheek off, and shredded his left tricep to ribbons. He was unconscious before his body slammed into the canal's mud with the thud of a thawed, wet turkey being dropped on a kitchen floor.

The invisible shockwave ripped the laces out of one of my boots and knocked me down. Sitting up, the first thing I heard through the ringing in my ears was Cavalier struggling to breath. It was an unnatural sort of wet cough, like someone was giving him the Heimlich while he lay face down in a puddle. I could tell someone was fighting hard to fill their lungs with air, but my ears weren't working well enough to tell who it was or where the strange sound was coming from. The smoke and dust clogged the air, making it impossible to see beyond my outstretched legs. I thanked God they were both still there.

I didn't know what had happened. My head pounded. I couldn't see anyone else. All I could hear was ringing and Cavalier's occasional labored breath. Broome emerged from the brown cloud I was sitting in and pulled me to my feet. He put his face against the side of my head and yelled, "That was Cavalier!"

It took my brain a few seconds to comprehend what he was saying: Cavalier had triggered an IED. The explosion seemed so massive I was sure multiple people were dead. All the previous IEDs I'd seen up close were firecrackers compared to this. Even the explosion that sent Rory's leg soaring through the air and over a compound wall had felt relatively small. This explosion throttled the Earth.

I followed Broome in a drunken stumble. I was sure each step I took was going to trigger another, even bigger IED. Every time our squad found one, there were at least two or more daisy-chained nearby. My mouth was stuck in a grimace, a psycho clown's dirt-filled grin. Getting off the road and into the field next to it felt like the most important task in the world.

Once our feet were off the hard-packed dirt and standing in freshly irrigated soil, we found Poth. He had a wet stain the size and shape of a pancake on his thigh. He noticed me staring at it.

"I'm fine, dude. A knee hit me," he said.

What he said didn't click, and he noticed that, too.

"Someone's knee flew past you guys and hit me in the leg."

Aside from a purple bruise that lingered for a week, Poth was fine. He had been standing about 10 feet behind Broome who'd been a few feet behind me when the IED went off.

By the time Poth explained the pancake to me, our corpsman and two other Marines had pulled Cavalier out of the canal he'd landed in and were working on saving his life. While they raced to cinch tourniquets on his arm and stuff his shattered pelvis with gauze, a dog handler who was attached to our squad asked for help.

"His leg landed in that cornfield over there. I need someone with a metal detector to help me find it."

I patted my side and found the plastic piece of gear was still there, so I joined him. Suddenly, finding Cavalier's leg became the only thing that mattered. I knew it wasn't going to be reattached; in fact, I was sure Cavalier was dead already. No one could survive something like that. But I didn't want the Taliban to take his leg as some sort of war trophy.

That leg belonged to a Marine, and they sure as fuck weren't going to put their hands on it.

Stumbling through chest-high corn stalks made using a metal detector virtually impossible, but I still swung it uselessly back and forth in front of my feet. I waded through the corn and tried to process what I was doing. Before I had a chance to grasp the fact that, at 19 years old, while most of my friends were hungover from spring break, I was in the middle of a minefield looking for my buddy's severed leg.

The dog handler found it.

He emerged from the corn carrying it by the ankle. Above the boot was an inch of white sock. Above that was several inches of pale, hairy skin I recognized as Cavalier's. Beyond that, the skin turned to flaps, exposing pink muscle and a piece of protruding ivory shin bone. My mind took a photograph. I was surprised there was none of the yellow fat I'd come to expect from other bodies I'd seen reduced to wreckage.

That photograph was on full display in my brain as I thought about how much more six hours of post sucked than four. I lit cigarette number two and put the photo back in the mental seabag with all the other things I wouldn't revisit until I got back home.

Later, I learned how the brain releases a tidal wave of endogenous adrenaline whenever it witnesses something as unnatural as a friend's athletic body split into pieces. The rush of adrenaline imprints the memory onto the amygdala, where it stays in the limbic system, waiting to resurface without warning. For me, that came after we returned home, while shoe shopping.

Wandering through the aisles, I was suddenly face to face with a mannequin showing off a new pair of Vans. The mannequin wasn't a full person, just the lower half of a leg. Its pale plastic contrasted with the dark green skate shoes and white laces tied in a double knot. I was fully aware I was looking at plastic, yet the sight of a lone leg wearing a shoe completely unnerved me. I broke into a cold sweat and fled the store, embarrassed and trying to catch my breath.

That reaction was more intense than what I'd experienced in Afghanistan. Even though I was fully aware there were no IEDs hidden in the racks of shoes, it felt like my life was in imminent danger. My legs took on that same jittery feeling as when I stumbled off the road after Broome. The cramped aisles and sudden confrontation with the plastic limb prompted what psychologists consider a normal response for short-term post-traumatic stress. The physical reaction — sweating, shortness of breath, and the overwhelming inclination to move — subsided after a few minutes. Those bodily responses were rare for me and disappeared altogether after a few years.

— >><< —

When the cigarette's cherry met the cotton filter, I flicked the butt on the dirt floor. I looked around the post and noticed Broome had left me a pack of Skittles. He was always trying to lift everyone's spirits. The first time either of us had been pinned down, he spent those eternal few minutes trying to make me laugh. We were stuck in the middle of a wheat field, and while we pressed our bellies and faces into the dirt, machine-gun rounds cut through the wheat inches above our heads like an invisible scythe. We couldn't see each other but were close enough to talk.

"Damn! We're in a tight spot!" he yelled, doing his best *O Brother, Where Art Thou?* impression.

It broke the tension and got us both moving. When we returned to the outpost that night, we sat on his cot and watched the movie on my iPod, one earbud for each of us.

I set the Skittles next to the 240 and cracked a water bottle. Drinking water and peeing it back into a bottle was one of the best ways to pass time on post. Depending on how much and how quickly I drank, the time it took to go through my body varied. Add temperature and time of day, and I could run a long list of water-to-piss experiments. I also learned that, if I held it until the last possible instant, I peed more. My goal was to break the record of how many bottles I could fill with a single relief of the bladder. If I got bored of the science experiment, I could see how many cigarettes I could smoke before needing Skittles to stave off nicotine sickness.

I was on bottle four and cigarette six when I felt something tickle my calf. I swatted at it without much thought. A few minutes later, the same tickle. I swatted again and looked down for a cause: nothing. *Here we go,* I thought. *Already starting to hallucinate.* My stomach, now distended and pressing painfully against my body armor, hated me. To take my mind off the absurd amount of water in my body, I lit another cigarette.

With the sun fully behind the horizon, I uncapped bottle five and took a sip. At that, a little piss leaked out of me like water through a crack in the Banqiao Dam. I reached for one of the empty bottles and frantically lined up the opening.

With my head back in ecstasy and a forceful flow of warm liquid bursting out of me, I was temporarily oblivious to the sanity-bending monotony of standing post.

Bottle two filled at a slower pace, and I moved on to bottle three before the stream declined to a trickle, then stopped. I capped the last bottle and set it next to the 240 with its comrades. I collapsed onto my throne and admired the results of the record-breaking piss.

Despite a successful science experiment in the limits of the human bladder, I still had several hours of post left. I was out of water, out of piss, and not in the mood for more cigarettes. The only remaining activity was to masturbate. With that thought, I felt the tickle on my calf again, followed by a tickle on my lap.

The Graveyard of Empires had been home to virtually endless war since the start of recorded history, and there was no shortage of stories about how haunted Afghanistan was. Marines from the previous deployment had sworn up and down that Gharmsir had been crawling with ghosts. Mysterious Russian voices were heard late at night, moving shadows were seen with no owners, and there was the ever-present sense that someone, or something, was looking over my shoulder whenever I stood post at night.

DOUBLE KNOT

I never believed the ghost stories, but the week before, a Marine in 3rd squad had nearly pissed himself while standing post in the middle of the night. He swore up and down that he'd heard footsteps stomping toward him. They had gotten louder and louder, as if whoever it was were walking at a brisk pace. Despite a bright moon, he had seen no one through his night vision, and based on the sound of the footsteps, there should have been someone directly in front of him. Just as the footsteps sounded practically on top of the post, they stopped. The Marine had been clutching his rifle and staring wide-eyed into the green-lit darkness of his night vision; no one was there. He had just exhaled and set his rifle down when the night's silence was shattered by a scream, mere feet away. He looked through the machine gun's thermal optic and saw the detailed heat signature of a man without feet sprinting away from the post. He watched in horror as the man emitting the heat signature never appeared through his night vision.

The story popped into my mind at the tickle on my calf. It made me nervous, but I had also just planted the seed of rubbing one out in my 19-year-old brain, so it was a toss-up between succumbing to fear or succumbing to hormones. *What's that ghost going to do anyway?* I thought. *Yell and run away?* With that, I said fuck it, stood up, and started unbuttoning my fly. Worst-case scenario, a ghost yells at me, probably inducing a fear-gasm. Best-case scenario, the ghost offers its services, and I will be the only Marine in the Corps to successfully solicit a handjob from a spirit.

Before I had the buttons undone, the lieutenant came bursting through the burlap sheet. He had somehow ninja'd up the wooden ladder without making a sound. With my attention split between my hand and what lay beyond the machine gun, I hadn't heard him.

I froze. But with more luck than it took to avoid an IED, the lieutenant did not turn toward me. He looked straight out into the darkness beyond the post. "How's it going? Anything unusual?"

Aside from the fact I was about to have my dick out and there might be ghosts in here, no, nothing unusual, I thought.

"Uhh ... no sir. Pretty quiet," I said. The pitch-blackness hid the situation in my pants.

"Very well. Carry on." And with that, he exited through the burlap and descended the ladder. I stood there in shock for several moments. I was embarrassed I had let someone sneak onto the post without hearing them, but the Lieutenant's lack of awareness was worse than mine. After all, I had kept watch on everything in front of me, which is where the Taliban and the ghosts tended to hang out.

I lit another cigarette, and my mind went back to the day Cavalier was hit. It was the third day of a company-sized operation, and we were on our way back. We'd been inserted at night by helicopter, but because we were grunts without the word "special" in front of our job titles, we were once again walking back.

I usually walked point when we patrolled as a squad, but this operation was bloated with people, and our positions got shuffled around. An hour before Cavalier stepped on the IED, I got into an argument with Poth. We were yelling at each other over which one of us was walking too fast or too slow, but really, we were just exhausted, frustrated, and sick of doing "the endless dance of the infantry." While we bickered like children, Cavalier quietly passed by and took point. I didn't take point again until Cavalier was already unconscious and on his way to a hospital in Bagram.

With 20 minutes left of post, I felt the ghostly tickle go from my ankle, up my leg, and onto my chest. I swatted at the feeling again and felt a crunch this time. Jumping up, I turned on my headlamp and shone it at the dirt. There in the dust, illuminated in red light, was the culprit behind the ghostly touches. Twitching in a pool of his own blood lay Jerry: a fat, dying mouse.

I turned the headlamp off and stared through my night vision at the same patch of Helmand Province I'd grown used to. Even then, in 2011, the landscape conveyed a sense of timelessness. This patch of southern Afghanistan had looked exactly the same in the time of Alexander the Great. It looked the same in the time of Jesus. It was going to look like this long after I, and everyone I'd ever meet, was dead.

Dead Weight

> *"It is not our part to master all the tides of the world, but to do what is in us for the succour of those years wherein we are set, uprooting the evil in the fields that we know, so that those who live after may have clean earth to till. What weather they shall have is not ours to rule."*
>
> *The Return of the King,* JRR Tolkien

WE MOVED IN three diamond-shaped formations. Four Marines per diamond, leap-frogging past each other. This allowed us to spread out and cover a substantial chunk of the Afghan landscape everywhere we went. It also allowed at least four Marines to watch for movement while the others could focus on their feet. IEDs saturated the poppy-rich countryside, and we were far more likely to die via an errant footstep than in the opening burst of an ambush. If one fireteam got attacked, the other two could maneuver.

By May 2011, midday temperatures stayed in the 90s, occasionally breaching 100. With 100 pounds of gear on our backs, those temperatures made everything a workout. Our squad hadn't hit an IED since Cavalier lost his legs two weeks earlier, but playing the Helmand leg lottery meant it was just a matter of time until an unlucky foot found another one. With each uneventful patrol, my anxiety grew.

I was sweeping an area for us to jump across a canal when the news came over the radio.

"Bin Laden is dead," Walsh shouted across the poppy field separating us. "Some SEALs killed him."

Broome, who was still waist-deep in the sea of pink flowers and sticky green stems between me and Walsh, turned back toward me and relayed the message.

"Yeah, I heard. Does that mean we get to go home?" I shouted back.

Broome laughed. We kept pushing.

I turned my attention back to my feet and resumed sweeping for IEDs. It seemed surreal that the mastermind behind 9/11, the bastard whose fault it was we were over here tip-toeing through minefields and killing people who, like us, were mostly children when the towers fell, was finally dead. I thought about how pointless it was to be walking through the Helmand countryside, anticipating an ambush or an IED going off, now that he was dead. What was the reasoning now? Kill every last Taliban fighter? That was a good enough reason to stop my mind from wandering any further down the rabbit hole, but not good enough to justify the next few months of killing — or the next 10 years.

A few weeks after we got the news Bin Laden was dead, things heated up in Trek Nawa. We found fewer IEDs but got in gunfights any time we patrolled north. A few kilometers south, the rest of the platoon set up a new patrol base — PB Lambert. They found multiple IEDs every day. That small speck of Afghanistan got so many IEDs that a detachment of EOD Marines was sent to live at Lambert. They told us that the area around Lambert now had a bigger IED threat than the notoriously deadly Sangin district. The previous spring, another battalion had lost 25 Marines there — the vast majority from IEDs.

While the rest of the platoon tried to avoid losing their legs at Lambert, our squad enjoyed regular firefights. If we patrolled south, west, or east, we rarely got shot at. If we went north, we were guaranteed to get into a fight. We decided to leave the cumbersome THOR IED jammer behind so we could carry extra ammunition. I filled my dump pouch with extra 40mm grenades. One of the SAW gunners started carrying an M240 and a backpack of belted ammo. We would simply walk north, baiting the Taliban to shoot at us, then trying to kill them all before they could get away. Sometimes, the Taliban would stick around and slug it out; other times, they would shoot, then disappear. By May, we didn't really care anymore. Firefights were fun, so long as Marines didn't die. It beat tip-toeing through the places where IEDs were seemingly everywhere.

At Lambert, things were different. The Taliban didn't care that Bin Laden was dead, and they still wanted to kill Marines. The Marines didn't care that Bin Laden was dead and still wanted to kill the assholes trying to kill them. The platoon still went on patrol, hit IEDs, and got shot at. The war remained neutral, casually collecting lives, indifferent to good and evil.

Matt Richard was killed one month after Bin Laden died. He was on his second deployment to Helmand, making him one of the more experienced Marines in the platoon. When the patrol's pointman found something metallic buried in the ground, Matt told the younger Marine to stay back while he investigated. It was too late in the deployment for one of his guys to get blown up, so Matt walked up to the spot in question and crouched down. He took out his Kabar and started probing the ground for signs of an IED. It detonated directly into his face.

The explosion sent his body flying through the air with one less hand, a missing jaw, and a broken neck. He was dead before his body landed on the Marine behind him.

The war didn't care who died or when. It just carried on the way it always had.

The war didn't just drag on after Bin Laden's body was dumped in the Arabian Sea; it raged on. His death came at the peak of the war's violence. He was killed on May 2, 2011, and by the end of the month, 56 more Americans were dead. By the end of the year, that number jumped to 563. Americans continued to fight in Afghanistan for another decade, seizing terrain, then handing it back to the Afghan Army, which invariably handed it back to the Taliban. This futile game of seize-and-relinquish offered no tangible benefits and cost another 1,329 American lives. In spite of the surging cost in souls, policymakers refused to admit what every soldier, sailor, airman, and Marine on the ground already knew: We were fighting an unwinnable war.

DEAD WEIGHT

Two years later, I was back in Afghanistan, sitting on a sprawling 16,000-acre base complete with its own roads, speed limits, and police to enforce them. Camp Leatherneck, which we dubbed "Camp Cupcake," was the most disgustingly swollen base one can imagine. Despite being the home to more than 2,200 Marines and sailors of the 1st Marine Expeditionary Force, it more closely resembled Minnesota's Mall of America. The base boasted air-conditioned trailers, hot showers, flushable toilets, fully-stocked gyms, chow halls that served steak and lobster weekly, a Pizza Hut, entertainment centers with TVs, computers, and guitars, full laundry service, and finally, the cherry on top, Green Beans Coffee shops.

At first, these were welcomed luxuries to those of us who had expected another seven months of MREs, burning our own shit, and sleeping next to our waste's eternal flame. But after a few weeks of living like a pig, my appreciation faded to disgust.

We were 12 years into the war, and America had its priorities completely backward. Fat American soldiers and contractors shuffled from watching TV in a recreation center back to their air-conditioned hooch with a cold drink in one hand and a Snickers in the other. Filthy rifles hung nonchalantly across their backs so as not to bounce against the guts spilling from their food-spattered camouflage. At some point, America had lost its way. The war's objective had seemingly shifted from winning to providing comfort to the American GI.

All these amenities existed because America had the deep pockets to fund the war it started, but no clear place to spend it. The gluttonous mustard stain of a base only worked to comfort the people living on it — all of whom were collecting that wonderful untaxed hazard pay. The fact that there was a greater hazard of developing diabetes than being hit by a rocket attack didn't matter. Those who had experienced the ugly realities of war couldn't help but feel complicit in this big fat waste of resources.

My time aboard Camp Cupcake wasn't completely without incident. While it was a night-and-day difference from my first deployment, there were reminders that a war, in some form or another, was still being waged in Afghanistan.

On one particular mission, we were sent to aid a platoon of Army engineers in a neighboring province. At the time, our company was acting as a Tactical Recovery of Aircraft and Personnel unit for the region. This meant one platoon at a time stood by with enough gear and ammo to hypothetically recover a downed aircraft or its crew from any conceivable jam. During one of our platoon's rotations, the engineers asked for some help in the mountains. There were no downed aircraft, but their convoy had been attacked, and a soldier had been killed. In the ensuing firefight, they were unable to retrieve his body.

They were attacked along a stretch of highway that took a sharp, almost 90-degree turn. The bend formed a natural bottleneck that someone with stars on their collar named the Devil's Elbow. It was flanked by steep mountains on one side and desert on the other. In an effort to thwart the ambush, the missing soldier had dismounted and led an assault up the boulder-strewn mountainside. After killing four Taliban fighters, he was shot in the face near the summit. His fellow soldiers had fought all night to keep the remaining Taliban away from his body, but they were unable to reach him.

Prior to our arrival, the Taliban disengaged and fled. We simply accompanied the soldiers up the mountain to their fallen comrade and helped carry him back down the rocky terrain. We found his body just below the summit, directly across the draw from two of the four fighters he'd killed. The position of his body, impossibly high up the rocky cliff at the end of a long trail of brass casings, spoke volumes of his bravery. It left me with a lasting respect for American soldiers. This man was not a Ranger or a Green Beret — nothing supposedly "special" about him. He wasn't even a grunt. He was just a soldier, determined to win the final fight of his life. He was a warrior to the last yard.

We put him on a body-sized unstructured tarp with handles called a poleless litter. At first, his comrades refused to let us help carry him, wanting to do it themselves. Eventually the descent became too arduous, and they relented. I grabbed one of the litter's canvas loop handles near the dead soldier's shoulder.

The climb down the slope was steep, and with every stumble, his head bounced against my leg. Despite the bullet wound below his left eye, his face and head were more intact than I expected. Excluding the small hole, he looked like he was sleeping, as if he might wake up and ask who I was. I looked away from his half closed eyes and wished I could apologize for the bumpy descent. Every so often, one of us would trip, sending the litter tumbling to the rocky ground. We did our best to avoid the sharp scrub brush and boulders, but it was impossible to climb down gracefully. It took eight of us to get him back to the Army's vehicles. The martian surface of loose shale and sharp rocks broken up by the occasional boulder gave me an appreciation for the soldiers fighting outside Helmand. As much as I detested the heat, sucking mud, and constant state of wet feet, I was glad to have fought in the farmland and poppyfields of the south.

I didn't understand *dead weight* until that mission. Practicing buddy carries and drags in training mimics the experience of carrying another human but falls short of replicating carrying a lifeless body. The smallest efforts of your partner eliminate any trace of reality from having to carry someone who is actually dead. With no help from the deceased, dead weight makes a small man much heavier than a large man whose heart is still beating.

Few people back home ever heard about that soldier's sacrifice, or the 162 other Americans who died in 2013. The nation continued to waste money and blood. Even then, I knew reaching that level of national nausea would take longer than it had during previous wars. Unlike during the draft-fueled Vietnam War, Americans now seemed indifferent about their military's fight on foreign soil. They wanted it to end but didn't really care how long it would take the all-volunteer force to throw in the towel. The war just wasn't close enough to impact their lives.

— >><< —

The war went on for eight more years. Knowing the outcome in advance did not assuage the sting of watching America flee in total defeat in August of 2021. I was glad to see a president finally have the gall to pull the plug on our longest war, but the manner in which Biden rushed for the door was unforgivable.

In April of 2021, he announced a complete withdrawal by the 20th anniversary of 9/11, directly causing a political and military collapse in Afghanistan. Even with four months before the total cave-in, the Biden administration did nothing to expedite the evacuation of our allies, forever damaging our credibility with future allies. His administration's desire to wrap up two decades of war with a nice 9/11-themed bow resulted in a botched exit that unnecessarily killed 13 American service members and nearly 200 Afghans.

Each of the four US presidents who held office during the war made unforgivable errors. In 2003, President Bush decided to invade Iraq, diverting American forces from the righteous war in Afghanistan to the criminal war in Iraq. This gave the Taliban enough breathing room to regroup in Pakistan. Two years later, the Talbian began an insurgency and successfully re-established themselves in Afghanistan. In 2009, President Obama announced a surge of an additional 30,000 troops (a number that would swell to 150,000 in two years) into Afghanistan to wrest the country free of the Taliban's grip, again. But in the same speech, Obama promised to begin withdrawing those same shock troops by the summer of 2011, completely undermining their efforts. His surge was doomed before it began. Then, in February of 2020, President Trump signed a back-alley deal with the Taliban, releasing 5,000 Taliban prisoners without so much as consulting the Afghan government. This signaled to the Taliban that the United States would recognize the Taliban as a legitimate party — even more legitimate than the Afghan government the United States had helped to establish.

So, while the blood of the last Americans to die in Afghanistan stains Biden's hands, blame for the other 2,448 dead Americans — not to mention the nearly 70,000 dead civilians and 1,100 NATO allies — falls on the three men who sat in the Oval Office before him.

I was in Virginia Beach when the United States finally tucked tail and scampered out of there. I was writing for a military news magazine at the time and doing my best to separate my personal feelings from the news I was reporting. I wrote about how the Afghan Army we trained and fought alongside dropped their weapons and surrendered without so much as a shot fired. I wrote about how the ancient citadel of Herat — the same fortress that withstood the advance of the world's greatest empires — got handed over to men in flip flops carrying old AK-47s. I wrote about civilians falling to their deaths from the wheels of American C-17s. I wrote about the last 11 Marines to die.

At Hamid Karzai International Airport's Abbey Gate, 11 Marines — alongside a Navy corpsman and an Army paratrooper — used their bodies to make a wall at the end of the world. While everything around them devolved into mayhem, the wall never wavered.

As I wrote their obituaries, I thought about the sickly sweet taste of over-sugared Green Bean Coffee. I thought about Matt, crouched over the IED, probing with his knife. I wondered whether he'd known how close he was to death and had poked at the deadly device just so his friends wouldn't have to. I thought about Christopher Meis, a machine gunner in my company, gasping for air with a hole in his chest. A hole from a bullet he'd been running toward when it entered him an inch above his body armor and ricocheted through his 20-year-old body. I thought about the six other Marines I'd served with in Afghanistan who'd died fighting the Taliban. I thought about the soldier I'd never met alive but whose lifeless body I helped carry off that nameless mountain. I wondered whether he'd known that his desperate stand would amount to little in the long run and that only a handful of his fellow soldiers would ever know his bravery equaled that of the warriors at Thermopylae and Normandy. And I wondered whether, at the moment their bodies were blown into eternity, the men and women who made the wall had been surprised that, in 2021, the war could still require dying.

Harden Up

"When the war is over, what will the soldiers do? They'll be walkin' around with a leg and a half, and the slackers, they'll have two."

Salonika, Ronnie Drew

IN MAY 2020, I got the news that Rory had killed himself. I was nine months into a new career as a cop and sitting in my police car behind an abandoned grocery store. It was that time of night when anyone wandering around was probably intoxicated or up to no good. The tall security fence behind my car meant I only had to keep an eye on the alley in front of me. The perfect place to sit quietly and chip away at overdue reports. It also turned out to be a good place to wrap my head around the death of another Marine I'd served with.

The news hit with the now-familiar feelings of surrealism and sadness. *Was he really dead? He sent me a text last week. He can't be dead.* But of course, he was. Just like the growing list of veterans caught up in whatever this invisible 22-A-Day killer was.

Rory had amassed a large following on social media. He seemed to have found a purpose in life after war as an advocate for awareness of the mental health struggles a lot of veterans go through. He'd capitalized on the fact that he was a wounded veteran who battled depression and PTS and had built an extensive network of friends on the internet. I couldn't put my finger on why, but his social media persona always felt disingenuous. I knew Rory at his best and at his worst from serving together in combat, and with each mental health post he made on Instagram, I grew increasingly worried. When he finally killed himself, it was the worst case of *See? I told you he was full of shit.* There was a reason he wasn't close with the Marines he'd served with.

I finished my shift, went home, and lay awake in bed. I wasn't particularly close with Rory. For most of the 10 years that I knew him, he'd been an asshole. I saw through the veil he hid behind but still felt obligated to let people know what Rory had been like at his strongest. I'd witnessed his bravery firsthand.

I reached out to a mutual friend with a substantial online platform and asked if he would share a story about Rory if I wrote it. He said yes.

I was with Rory when he stepped on the IED that blew his leg off. We had been in Afghanistan for only two weeks. The explosion was muffled. The Taliban had buried the IED too deep in the Helmand mud. The homemade bomb still had enough bite to rip Rory's leg off above the knee. As I turned around, I watched it fly over a 10-foot compound wall and land in the muddy field I was lying in.

Memories are strange and unreliable, but I know that, in the next moments, in some order, Marines applied tourniquets to Rory's thighs, swept the compound for secondary IEDs, cleared a landing zone, and called a medevac helicopter. What sticks out in my mind is how Rory acted while he waited to be evacuated.

Once our corpsman stabilized Rory, and that eternity began, Rory calmly asked for a cigarette. He lit it and joked with the Marines and sailor huddled around him. Cigarette smoke replaced the lingering chemical smell left by the explosion. He was fully aware that one of his legs was gone. The amount of blood on those who'd come to his aid confirmed his life hung in the balance. There are few instances in combat when losing one's composure is acceptable. This was one, but Rory maintained his usual calm demeanor. The fear and pain were there, but he buried them for our sake.

It was Rory's third combat deployment. His second to Afghanistan since Obama's surge. Most of the squad was on their first deployment, and this was the first time most of us had seen someone get blown up.

There was a viral video going around at the time of an Australian guy with aviator sunglasses and a big mustache yelling at people to "Harden the Fuck Up!" Rory bought a patch with that guy's face and wore it on his body armor. When one of Rory's younger Marines seemed frazzled by what was going on, Rory smiled. He took a drag from the cigarette and slapped the patch into the Marine's hand.

In the story, I wanted to pay my respects. I wrote about how Rory's stoicism set the tone for the next seven months. His courage strengthened our squad at a critical moment, early in the deployment when most of us were inexperienced and impressionable. I wanted people to know how strong he was. But as time passed since Rory's death and more Marines killed themselves, I started to feel differently.

Another Rory I served with, Rory Dalgliesh, shot himself in 2013. We'd gone through a dog handler training course together. He had been one of the funniest Marines I knew.

Two years later, Justin Walker, a member of my platoon who once volunteered to steal six cakes from a supply truck when we were without MREs, went to a shooting range, rented a rifle, and shot himself in the heart. Brian Terra, a well-liked machine gunner, also took his own life. So did Daniel Salter, Paul Oliver, Michael Harris, Adam Wolfel, Jared Paynter ... the list goes on.

There was a time when suicide was taboo. It used to be condemned as an act of cowardice. Now the act has become so ordinary that veterans think of killing themselves as a viable option for postwar life, wedged somewhere between entrepreneurship and homelessness. The obvious cons of such an irrevocable choice (inflicting permanent pain on those you love and, of course, death) are obscured by blustery rhetoric and mythmaking. Now, when a veteran commits suicide, an army of former comrades and virtual acquaintances will inevitably hoist him up alongside service members killed in combat, just as I had initially done with Rory following his death. Some people have even taken to wearing the names of those who kill themselves on memorial bracelets — a practice once reserved for comrades killed in battle. Droves of people share pictures of the deceased during their glory days, wax poetic about seeing them again in Valhalla, and gush about how their decision to commit suicide was tragic yet somehow understandable.

In our efforts to stop veterans from dying by their own hands, we have worsened the problem. Pushups and T-shirt lines have loosened the proverbial tourniquet and released the severed artery of self-inflicted gunshot wounds.

I've lost more than one of my personal heroes to suicide, and criticizing them feels wrong. However, I have come to realize that it's necessary. One month before I left the Marine Corps, my grandfather, a career Navy veteran and patient patriarch of the family, shot himself in the temple with a revolver. He lived just long enough to stumble from the garage and collapse on his front lawn. Following his death, all I wanted to do was sing his praises and ensure the rest of the world knew how great a person we lost. It's now clear that this is a disservice and only raises the likelihood of someone else's hero choosing the same irreversible path.

As distance grew between those of us who fought together in Afghanistan and our shared time there, contact tapered off. Frequent calls and text messages dwindled to annual check-ins, which eventually faded to nothing.

When Broome died, it had been more than two years since the last time he'd responded to one of my texts. Broome's death did not come from the barrel of a gun but from the mouth of a bottle. One year before he died of liver failure, his doctor told him his drinking habits would kill him in a matter of months if he didn't stop. Broome's mother reached out to Poth and begged him to fly to South Carolina in a last-ditch effort to stop Broome's self-destructive downward spiral. Poth later told me that Broome had aged 25 years in the 24 months since they'd last been together — gaunt and with a yellow face so covered in wrinkles he was hardly recognizable. Broome refused to listen to Poth's pleas and kept on drinking. He knew exactly where the train was headed, and he made no attempt to get off.

Broome, the man who was equally fearless in combat and gentle in friendship, died a painful self-induced death. If I could go back and have one more conversation with him, I'd tell him he owed me more time together.

He had been willing to die for me, and he had proved it over and over. Whenever I got nervous about crossing canals too large to sweep the far bank for IEDs, Broome would often run and jump across first. He would assume the high risk of landing on an IED just so that I wouldn't have to. He took more than his share of the load whenever he could.

The world is colder without Broome in it. His death stole something beautiful away from everyone who knew him, leaving behind more pain for others to shoulder. I like to think that, if he had only realized what his death would rob us of, he might have fought harder to stick around. Of course, he'll never help those he loved share another burden. His choice — like Rory's and all those who've made it — is permanent.

When Pigs Fly

"It's good to be good, but it's better to be lucky."

Homicide: A Year on the Killing Streets, David Simon

LATE IN THE afternoon of March 23, 2020, Maryland Governor Larry Hogan ordered all nonessential businesses to close their doors in what felt like a too-little-too-late attempt to prevent the spread of COVID-19. Less than two weeks earlier, on March 11, the first quarantine area was established in New Rochelle, New York. Five days later, San Francisco announced the first "shelter-in-place" in the country. The wait was over. The 21st-century plague had crossed the Pacific, landed on American soil, and was sprinting across the country like stage four lung cancer.

No one knew how to handle it yet. The experts had more questions than answers: How does it spread? Is it airborne? Do masks help? Can you contract the disease from bodies of those killed by it?

There were no definitive answers, and the beginning of the Great Vaccination Divide was still nine months away. All the uncertainty led to fear. Lots of fear. Everyone was afraid of everyone.

As the COVID body count started to skyrocket and mass graves were being dug on New York's Hart Island, the ordinary duty of first responders to gather the dead felt like a death sentence. I was a rookie cop at the time — three months on the job — and responding to calls for the deceased quickly started to feel like playing Russian roulette with Agent Orange.

On one particularly slow night, a caregiver called the police when she couldn't get ahold of the elderly woman she looked after. She'd dropped groceries off earlier in the day but had left the two bags of food by the back door when no one answered. She figured the elderly woman was taking a nap, but that was six hours ago, and she still couldn't reach the old woman on the phone.

When I pulled up to the house, I knew whatever I was about to find wouldn't be good. It was an old Victorian house up on a hill surrounded by even older sycamore trees. Stephen King couldn't have dreamt up a better setting for a horror story. I had just finished reading *The Stand*, so the whole world looked doomed to me. The groceries were still by the back door, and the half-gallon of milk was now warm.

A second, more experienced cop showed up, and after neither of us could see anyone through the windows, we knew one of us had to go inside. However, the doors and windows were locked. We searched the property for a spare key but came up empty.

"Fuck it, man. I'll force the door and go in, but I think you should stay out here in case the house has COVID in it," he said.

I didn't need any convincing to stay outside the disease-ridden haunted house. I was in no hurry to go find the old lady's body and risk contracting the invisible killer.

"Go for it, man. Holler if you need something."

It wasn't hard to push past the aging door. The other cop went in while I meandered about and admired the old house's steeply pitched roof and ornate gables. He came back out a few minutes later and confirmed our suspicions.

"She's dead, sprawled out on the top of the stairs. Looks like she just kind of laid down and died," he said with all the passion of Ben Stein.

We went through the standard procedure: called medics to come pronounce her dead, notified the supervisor, then waited for the medical examiner to show up. We sat on the back step and watched the sun go down. Once it was dark, we could see the old lady's legs through a window, stiffening in their stockings. It looked as if she'd felt like lying down and just never got back up. Most Americans were exhausted then; she just seemed to have succumbed to it.

The call took a particularly long time because it was getting late, and shifts were about to change over. We were also in the middle of nowhere, and no one was in agreement about whose jurisdiction it was to handle the woman's body. With COVID on everyone's mind, no one wanted to do it. I thought the medics should, since they already went in and touched her. The medics said no way, they had other calls to handle. The medical examiner's crew normally removes bodies, but with COVID, they didn't want anything to do with it. I didn't blame them. At the time, dragging a COVID corpse down some stairs felt life-threatening, and they didn't get paid enough to gamble like that. Ultimately, the other cop stayed with the body until a reluctant worker from the medical examiner's office drove out with a van and took the cold amalgamation of elderly bones and flesh away.

My rookie brain had no point of reference for normal police work, but COVID made the job feel stranger than it needed to be. A precinct I didn't work in was dubbed the COVID precinct, and all prisoners had to be processed there. No big deal, except that I had to learn the ropes from people I didn't work with. They were generally annoyed to have to show the new guy how to do things. I spent a lot of late nights working well into the next shift at a precinct full of strangers. It was isolating.

Policing as a whole was lonelier than I had anticipated. You didn't ride around in your police car with a partner like in the movies. There weren't enough cops for that. You responded to and handled calls alone and spent about 90% of the shift alone. Someone backed you up if you requested it, and any downtime was spent parked together catching up on reports, but overall, policing was a solitary job.

Unlike most cops, who feel called to wear a badge, I sort of fell into it. When my last semester of college started to wind down, I felt the pressure to start a career. My history degree wasn't going to land me any sweet gigs out the gate, and the only thing on my resume was Marine infantry. I'd studied social work before switching majors to history, and in my mind, cops were the halfway point between social worker and infantryman. On paper, it made sense.

When I applied, the fire department didn't respond fast enough, and the police department practically knocked my door down trying to get an able-bodied citizen to join. They offered decent pay, great benefits, and the potential for rewarding work. My desire to serve my community mixed with the fact I already knew I handled emergencies well made being a cop look like a pretty promising path.

I lasted a year.

The sixth-month police academy went well enough. After Marine boot camp, the screaming, overweight, and over-40 instructors didn't frazzle me. I became a squad leader, and after leading Marines in Afghanistan, leading soon-to-be-police officers through basic police training wasn't hard. I did well in academics, shot well on the range, and handled the mock-emergency calls to their standards. I was no super-recruit, but the academy left me feeling confident that police work would be an easy and satisfying path to a pension.

Then, on May 25, 2020, five months after I graduated, George Floyd was killed. Suddenly, the crowds who'd praised first responders as heroes in the early days of the pandemic changed their tune. Anyone with a badge was just a racist with a gun.

I have thick skin, and most of the noise bounced off me. I still encountered grateful citizens every shift, but more people began hating us just for being cops. The constant bombardment from friends and family asking me how I felt about Floyd's murder — or any police incident that made the news — is what really exhausted me. *Did you hear about that shooting in Georgia? Why didn't they just shoot them in the leg? What would you have done?*

Fielding questions on behalf of cops across the country was draining. I rarely knew any details about whatever incident some well-meaning friend wanted my not-so-expert opinion on. I was still trying to learn the job — figuring out how to help people and lead with compassion, while simultaneously staying safe enough to go home to my family every night.

With every platform on my phone feeding me videos of cops fucking up and the protests that followed, I felt like I was on the wrong side of history. Even though I had a front-row seat to modern policing and selfless service from the cops I worked with, I couldn't help but worry that my daughters would grow up and see me as no different from cops in the 1960s. Unfortunately, our memories mostly play images of police K9s terrorizing activists and firehoses blasting Americans off sidewalks. I didn't want to leave behind a similar legacy.

With a steady flow of hey-aren't-those-cops-racist questions from my friends, I started to approach each call for service with *make sure people know you're one of the good guys* echoing in my head. But that voice got too loud, and I started to neglect safety.

On one particular call, I helped look for a suspect who'd recently kicked in someone's apartment door, hit them with a pistol, and robbed them. The description was only a young black male, armed and dangerous. While walking through the apartment's parking lot, I came face to face with a teenager who fit the description. My hand went to my gun. I was trained to draw my weapon if there was even a chance I might be face to face with an armed suspect. But that little voice spit in my ear, *That's racist! This poor kid is going for a walk*. I let him walk right up to me, and after questioning him for a few minutes, I handed him off to another officer and kept looking for the suspect.

A week later, I found out that the poor kid who I'd decided was just going for a walk was the pistol-whipping robber. He told detectives, "If that skinny bald cop had come around the corner 30 seconds earlier, I'd have killed his ass." Evidently, he had thrown the pistol in the gutter just before we'd found each other. By sheer luck, my skinny bald ass lived another day.

I took that lesson to heart and stopped worrying about perception as much. I knew what was right and wrong, and I was going to stick to my gut as my guiding light. No more taking chances with armed suspects so as not to risk ruining someone's walk. A few weeks filled with satisfying calls for service passed, and I decided helping people was worth whatever unfair judgments were befalling cops. I intended to make a difference, but that changed when my boss tried to give me a medal for something I didn't do.

It was June, and the bipolar weather of Baltimore in the springtime had finally given way to summer. Everyone was eager to be outside, their bones fully defrosted. I was sitting in a parking garage, closing out reports and sipping the free coffee Wawa gave to cops. The half-empty paper cup was now completely without warmth. My long-silent radio crackled back to life.

"Three-one-one. Start me some more units. Those street racers are back."

Three-one-one is Kris today, I thought. *Those fucking street racers are going to get someone killed.*

I'd already dealt with this mob of car enthusiasts enough times to know nothing good ever happened when they were around. I swallowed the last of the brown fuel with a grimace and popped the department's piece-of-shit Ford Taurus into gear.

"Three-one-one. I need a medic and some more units," Kris said over the radio. "Someone got run over."

I flipped on the lights and sirens to release myself from the traffic. I could tell by his voice that whoever had been hit was in trouble. Kris had one of the coolest demeanors on the shift and rarely allowed emotion to sneak onto the radio.

"Three-one-one. Let medics know he's got a serious bleed in his groin, and there's —" At that, I kicked it into gear and started flying. My sirens didn't have quite the desired effect — like Moses had on the Red Sea — but most cars still moved out of the way. With each car that refused to get over, my blood pressure spiked. To compensate, rather than stopping and ensuring each intersection was clear, I slowed down just enough to convince myself I could cut through without getting T-boned. The Taurus shuddered with the high RPMs, and I couldn't help but smile as adrenaline overcame agitation.

I knew that Kris, along with almost the entire shift, had a comically low level of medical training. First aid in the academy consisted of CPR, Narcan administration, and tourniquet application. Anything beyond that was left to paramedics. In the Marines, I'd been lucky to receive some more training and tried my best to respond to other officers' calls that might require a little more than CPR.

The training had been run by special operations guys from the Army, Navy, and Air Force. I spent a week in the classroom learning battlefield medicine, from blast injuries and gunshots to how to properly administer an IV for serious hangovers. The second week focused on testing the newly acquired skills on one another, then on dummies with prosthetics, and finally on live tissue — typically pigs or goats. It was this last detail that made the course so valuable.

Animal rights activists were supposedly waging a constant war against any training that involved live tissue.

I'm an animal lover, but I struggle to understand how even PETA can't see past the dead animals to the truth — dead pigs save lives. An easy trade for any compassionate person.

The final day required the class to meet at a remote location in rural North Carolina in civilian clothes. After arriving at the training site, the other students and I gathered under a few tents surrounded by a privacy fence. A dozen pigs were laid out on the ground, each one sedated and monitored by a veterinarian. The instructors explained that students would pair up, and each duo would be assigned a pig. While we were being hazed via calisthenics, our "patients" would incur some kind of injury at the hands of the instructors meant to mimic battlefield wounds. Upon cries for "Corpsman up!" we would run to treat our patients. Once the pigs were stabilized, we were sent off again for more exercises, then called back to treat further injuries. This process continued until we failed to stabilize our pigs.

I was on burpee 30 when the first cries came.

"CORPSMAN!"

I sprang to my feet and rushed toward the pig. Its front leg was severed at the shoulder joint, and blood flowed out steadily, like a bottomless cup of fruit punch upended by a toddler at the dinner table. I pushed hard on the pig's leaky armpit, slowing the bleed, while I grabbed and then applied a tourniquet with my other hand. Despite pig anatomy being remarkably similar to that of humans, their limbs just aren't built for tourniquets. It worked to stop the bleed, but it was clear that as soon as I needed to move my patient, the cinch would slip, causing the pig to start bleeding out again. While my partner held the stump in place, I jammed as much hemostatic gauze as I could into the extremity, then capped it off with a compression bandage. It was ugly, but the bleeding stopped, and the pig was deemed stable.

After more burpees and monkey-fuckers, cries for a corpsman sounded again. This time the pig had a hole in its chest. Pink blood bubbled out of the hole with each breath, and the pig's trachea began to push to one side beneath its fatty neck. A chest seal and decompression needle temporarily fixed the problem before I was sent back to the tree line to sweat some more.

The wounds continued to add up, and soon, patients began to die. Finally, my pig was the last alive, though it barely resembled swine anymore. Half its limbs were gone, its jaw was smashed apart, and red bandages obscured the countless holes and lacerations on its chubby body. The plastic tubes protruding from its snout and throat made it look more like a monster born of a mad scientist.

Out of medical supplies, I answered the call for a corpsman a final time. The instructor had given the pig a fatal cut with a scalpel, a small laceration in its armpit, nicking the brachial artery. I forced my fingers in the hole up to the third knuckle. As the warm fluid spilled over my hands a final time, I looked around for help. It was just a pig, but I knew that, in just a few short months, it might be a Marine whose life was flowing through my fingers. I saw the severed hoof lying on the grass next to the pig like a discarded food scrap and picked it up. I jammed it into the armpit's opening and used duct tape to fix it in place, stemming the flow of blood. The pig's vitals stabilized.

I was given a fatherly pat on the back, and my pig was given a merciful dose of sedative. Two other students grabbed the pig by its remaining limbs and labored off toward the bed of a waiting pickup truck. On the obviously unpracticed count of three, they heaved the mutilated body into the air with enough force to land it on top of the heap. The flight ended abruptly with a wet smack, then the fat, hairless corpse slid down the pile of pink bodies to a sticky, red stop at the truck's tailgate.

Leaving that day, head-to-toe stained brown with dried blood, I closed my hands in prayer and lied that I hoped to never use the training in real life. The training was far from enough to make me a professional in battlefield medicine, but it filled me with dangerous confidence.

Years later, in a blue uniform, I still carried the misconception that I could treat most traumatic injuries. After all, my pig had lasted longer than the rest.

— >><< —

"Three-one-four, three-one-one applied a tourniquet, but the injury is in the groin." Kris came over the radio. "How far out are medics? We need more units here; I've got nosy onlookers up my ass."

Driving faster than I ever had on a road with mailboxes, "98" flashed on the speedometer. The passing trees were a green wall. Street racers flew equally fast in the opposite direction, leaving their friend to bleed out alone. They chose fear of repercussions over loyalty. I eased off the gas as I rounded a curve, revealing a world of flashing red and blue. I moved my car as close to Kris' as possible before there were too many people crowding the dimly lit road to continue.

"Three-two-six. I'm out," I said as I ran around the Taurus' hot hood to retrieve my personal trauma kit. I built it myself after opening the ill-equipped department-issued first-aid bag for the first time to find a single tourniquet, one roll of paper towels, and a bottle of hydrogen peroxide.

I pulled the door handle. Locked. *You fucking idiot*, I thought. *Slow the fuck down.* I stuffed my excitement down to where all guilty memories live, scurried back around, unlocked the car, and grabbed the canvas bag. As I started running toward the center of the asphalt stage, the epicenter of the growing commotion, Kris met me coming the other direction.

"Yo, I need something more —" His eyes caught on the kit. "Yeah, dude, something like that! His shit is fucked up too high for a tourniquet."

I got to the center of the small crowd and found Nik, another new cop on the shift, leaning over a man in his 20s, pressing a balled-up pair of pants into his groin. The denim was saturated. Apparently, the man had been on a motorcycle, and in the confusion created by the smoking tires of his buddies' donuts, he had inadvertently been run over. Now he lay on the ground, semiconscious and bleeding profusely. Scrambling to pull the impossibly small latex glove over my sweaty hand, I directed Kris, "Put your knee on his groin," I pointed. "Full weight."

Kris was a veteran cop but immediately followed my instruction. Like most people in extraordinary situations, he was eager for guidance. I finally got the glove on far enough, empty reservoirs still dangling from my fingertips. I pulled out a packet of Combat Gauze and ripped it open.

"All right, lift your knee so I can get in there."

Kris obliged, exposing a rip in the man's groin the size and shape of a lime. I started feeding the gauze in, as deep as my fingers could reach, massing the material against the hemorrhaging artery. The hole swallowed the gauze with ease. I felt as if I were filling a warm void with gauze, pinch by nauseating pinch.

"Nik, open his eyes and keep him awake; just talk to him."

Nik tried, but the man was somewhere else. His eyes looked at things no one else could see. He moaned but didn't know it. Miraculously, the hole started to fill, and when no more gauze could fit inside the wound, I balled up the rest and held it on top.

"Kris, hand me that red package there." Kris frantically grabbed the red package. "Open it first, please," I said, my hands still busy with the wound.

"Yep." Kris opened the package and took out the large bandage.

We switched roles so I could secure the compression bandage on top of the protruding gauze. I surprised myself with how comfortable I felt. My movements were fluid, my directions clear and calming. I felt at ease for the first time since swearing in. Somehow, this was less stressful than the paperwork that accompanies a shoplifter. When I had the bleeding stopped, I leaned back on my heels.

"Three-two-six. The bleed is stopped, but he needs to get to a hospital. Medics need to step it up." Before I had unkeyed his radio, a paramedic materialized. A brief exchange caught the paramedic up to speed, and the no-longer-moaning man was moved into the ambulance and on his way to being someone else's responsibility.

Try as I might to fight it, I felt grossly proud. I had saved a life and been dependable when others needed me. Months of being teased for carrying a trauma kit had paid off, and the training I'd neglected since the pig was once again brought to bear. *Fuck yeah.* I congratulated myself while outwardly faking humility. As soon as the man had been placed in the care of the medics, my thoughts turned to my own ego.

I spent the hours after shift fighting a disgusting smile. When I returned to work the next day, my head barely fit through the door. Walking into the report-writing room, I was ready to lie and say, "Oh, it was nothing. All in a day's work." I was starting to feel too damn good about myself when reality arrived right on schedule for an ego check.

"Hey Mac," my supervisor said as he stuck his balding head in the doorway, "that dude didn't make it. Not exactly sure why. They called it on the way to the hospital." He slapped the door frame and continued on his way down the hall to something more important. He certainly didn't want to miss an opportunity to laugh at LT's jokes.

I sat there alone in silence. The man was dead, his DNA was still being washed out of last night's uniform, and it didn't matter at all. I was already used to death; overdoses and car accidents were commonplace, but this guy had been alive. Everything I'd failed to do flooded my brain. I felt sick. In my haste to be the hero, I had prevented some blood from flowing onto the pavement, nothing more. I hadn't rolled the man over to check for injuries on his back. I hadn't treated for shock, or even gotten the man off the cold roadway. I hadn't secured his neck. In fact, I had done nothing beyond treat a single bleed. I had haphazardly completed step one of five, then concluded my work was done. I did just enough to absolve myself of shame for inaction. Before I could wrap my mind around it all, my supervisor's sweaty face popped back into view.

"Oh, I'm putting you in for a lifesaving medal though," he said. "I know he died, but you deserve it. Apparently that guy was a medic in the National Guard. Good kid. Anyway, roll call in five."

A burp echoed in the hall with his shuffling footsteps.

I had my exit interview a few weeks later with the lieutenant. It was strange to see my name on two seemingly contradictory papers sitting on the same desk: on the left, a letter of resignation; on the right, an award recommendation.

I was used to some medals being about as truthful as comic books. The military is full of people showered in brass for things they never did. I remember a Marine getting a Bronze Star for throwing a hand grenade into a chicken coop. We'd all had a good laugh about that one. When I pulled out of the precinct parking lot for the last time, I felt relieved. I would only have to bear the shame of failure without the added weight of a medal.

When I resigned, I had no idea what I was going to do for work. I had vague ideas of trying to be a writer but no idea where to begin. For fast cash, I went back to the Irish pub I had bartended in during college. Instead of slinging drinks, I took a job as a bouncer. My old coworkers welcomed me back with open arms. Part of me went crawling back to the same pub so quickly because I knew they would. One of them bought me a welcome-back gift. Another told me they were proud of me for "doing the right thing." I welcomed the love but knew I was full of shit. I didn't hate cops like they thought I did. I felt like a hypocrite. I was no different from a Vietnam veteran throwing his medals on the White House lawn to win the respect of new, more progressive friends.

The Dog and Pony Show

> *"Nobody said you ought to like it. Nobody pretended to like it. Everyone you met took it for granted that the whole thing was an odious necessity, a ghastly interruption of rational life. And that made all the difference... There were nasty people in the army; but memory fills those months with pleasant, transitory contacts. Every few days one seemed to meet a scholar, an original, a poet, a cheery buffoon, a raconteur, or at the least a man of good will."*
>
> *Surprised by Joy*, C.S. Lewis

Pensacola, Florida

SWALLOWING THE LAST bit of beer, I placed the empty pint glass on the bar of the open-air Florida brewery and looked at my watch — recently switched back to military time to avoid any chance of miscommunication in the coming week. 16:26. I still had more than an hour before The General and his entourage arrived at Naval Air Station Pensacola for the second stop of what I was calling the "Joint Chiefs' Whirlwind Tour of Military Bases." Two beers can quickly turn into too many when you're drinking to calm your nerves, so I ordered a water and settled up.

This was my second reporting assignment, and imposter syndrome was in full swing. Unlike the previous assignment, which required only one day of reporting and a travel day on each side, this trip entailed a week of steady dispatches. With enough luck, anyone can fake it for a day, but I was convinced a whole week would expose me as a fraud. The voice in my head persistently tried to convince me I was unqualified for the job. I was a newly minted reporter (*or was I a journalist?*) with no relevant degree or even an internship at a local paper. I had a blank resume when it came to professional writing. Shit, I didn't even have a legitimate camera, just my iPhone.

My plan was to lean into the fact I was a veteran ("See, guys? I'm one of you!") and charm their pants off. I pulled out the veteran identity I kept hanging in my closet, nestled between ex-cop and half-assed hippie. A couple deployments to Afghanistan gave me just enough street cred to be welcomed into the staff's inner circle. A little time under fire gave me a ticket to rub elbows with the tour staff rather than getting sent to the back of the bus, where I imagine they usually stuffed embedded reporters. In too many veteran circles, simply having experienced combat — even on the periphery, through something as random as having a few inaccurate rounds fired in your general direction — holds more weight than it should. Just being in the right place at the right time is often the price of admission into the cool club. As arbitrary a qualifier as it is, I shamelessly flashed my cheap credentials to make nice with the brass.

The tour was scheduled to hit seven bases in seven days, and I was tagging along for the whole thing. I'd live, eat, and sleep with *the talent* every step of the way. And by sleep, I mean rest for several hours a night, alone. It's important to clarify that, as the leader of this traveling roadshow — whom I'll call The General — had just emerged from some pretty scandalous accusations made by one of his former aides. (The General was found innocent, the accusations were dropped, and I was expected not to ask him or anyone on his staff about the recent rumors of impropriety.) My job was to report on all the great things The General, his army of staffers, and the motley crew of entertainers were doing for the troops. Nothing more.

I was one day behind The General and his posse because of a passport hiccup. They kicked off the tour with a stop in Guantanamo Bay, Cuba. Today, they were on their way to Florida, where I was to meet them for the second stop.

THE DOG AND PONY SHOW

My point of contact was a Marine Corps veteran whom I will call Niall. He was short, bald, and gave off a very squirrelly vibe. He wore a permanent smile, and his eyes were always stretched to the point that the white sclerae made complete circles around his irises. He existed on cocaine-level energy, but I can all but guarantee the public-affairs-officer-turned-USO-liaison never touched the stuff. He was one of those people who exuded such inexhaustible positivity that you questioned whether they lived on the same planet as the rest of us. Didn't he know about things like war and famine, or that real monsters like serial killers and child rapists existed? Hadn't he ever been dumped or had his heart broken? Of course, once you get to know people like him, you realize that, not only are the Nialls of the world not naive to the way of things, but more often than not, they have experienced tragedy on a level more intimate than most. By some stalwart faith — or maybe a sprinkle of magic positivity dust — the Nialls remain unjaded. When I finally saw his bald head on the other side of the airplane hangar, he immediately welcomed me with open arms.

He filled me in on the 24 hours I'd missed. The tour group — The General, his posse of high-ranking assistants, the two-man country band, the comedian, a DJ, Miss America, her personal minion, the USO staff, and, of course, the general's personal chef — had flown to Guantanamo Bay and met with the poor souls stationed off the coast of Cuba. The General had shaken hands and told the troops America was proud of them. The country duo played their two better-known songs, then played such guaranteed hits as *Friends in Low Places* and *Sweet Home Alabama*. Miss America looked beautiful and dutifully posed for photos with drooling 18-year-olds. And the chef made a delicious five-star meal for The General and his circus.

I quickly got the sense that this year's tour was a big deal. COVID restrictions were finally lifting, most people were getting vaccinated, and the USO was eager to get back out there and entertain the troops. The annual tour reached 32,350 troops the year prior to the lockdowns, and the USO aimed to beat that number after a year of virtual events. To accomplish this, there would be no downtime. We'd arrive at a base, set up, perform, meet the troops, then load back up on our C-130 and fly to the next base. A few stops allowed for short nights in hotels, but most of our sleeping would be done at 28,000 feet inside the airplane.

Before the Pensacola show got underway, Niall took me around and introduced me to the rest of the tour members. Lots of high-ranking officers and enlisted aides — nearly all of whose names I would have forgotten, had they not been wearing them on their chests. The country duo was clearly the life of the tour. With their faux-hawks, tight pants, and Axl Rose energy, everyone wanted to be wherever they were. I later learned that the USO tour was a paid gig, but the duo was adamant that they joined the circus because they wanted to support the troops. Neither of the rhinestone cowboys had served, but they were both quick to tell me about a cousin or an uncle or a half-brother who did.

That compulsion to immediately convince a veteran of some kind of military connection is a strange yet common phenomenon. I've never understood why people feel the need to prove that just because they didn't wear a uniform doesn't mean they don't appreciate those who did.

One of the uniformed people on the tour immediately stuck out from the rest of the camouflaged crew. I'll call him Raymond. He was the highest-ranking enlisted service member in the entire DOD, also known as a big deal. He'd spent the majority of his career in special operations and could kick the ass of anyone we met during the tour. He was in his 60s and in better shape than the rest of us — including the rockstars, the soldiers, and Miss America.

"Mac! Great to meet you," he said, sticking out his hand. "I'm sure Niall is happy to have another jarhead tagging along. Looking forward to talking."

I shook his hand and did my best to match his enthusiasm. Before Niall pulled me away to meet the next person, Raymond's smile turned serious. He leaned in to say something discreetly.

"I'm glad you're joining us. We gotta put an end to all this anti-war bullshit."

Miramar, Florida

The first show was genuinely fun. It had been a decade since I'd worn the single stripe of a private first class, and seeing how excited the privates in Pensacola were to do something besides drink in their barracks was contagious. It felt downright wholesome. Older service members brought their kids, who all waved those tiny American flags you see during 4th of July parades, and the junior enlisted all ran around like high school students at a freshman dance.

That night we slept in Pensacola before touring the base the following morning. I stayed up late typing up my first dispatch, editing photos, and thinking about Raymond's unprovoked anti-war comment.

Raymond assumed that, because I wrote for a pro-military magazine and had once been a grunt, I must be pro anything that involved American ordnance finding a home on foreign soil. He would've been right about Sgt. Caltrider, but 30-year-old fledgling writer Mac was not on the same page. I didn't have the heart to tell him that I thought the Iraq War was a mistake and the Afghanistan War should've ended years before I ever got there. I decided to keep my ever-evolving attitude toward war to myself. To get the most out of this assignment, I'd passively be whatever The General and the circus expected me to be.

The next morning, we ate breakfast in the chow hall and mingled with the troops. We ate the same eggs, sausage, and grits that chow halls served a decade earlier. Even the cheap Pizza Hut-style plastic cups hadn't changed. Most of the young sailors and airmen tried to avoid tables with generals. The country duo was surrounded, as was one of the *Super Trooper* actors who'd joined us. After eating our fill, we went to the Aviation Rescue Swimmer School: home to some of the Coast Guard's finest troops.

I frantically snapped photos of the Coast Guard's newest rescue swimmers and gathered quotes about the difficulty of their training and their eagerness to start conducting real-world rescues. Their collective mentality crushed any concerns I'd had that our military was getting soft. These kids were in their prime and wanted nothing more than to do their jobs. They were the best in the business.

Before we left, one of the cocky new Coasties made the mistake of challenging Raymond to a pull-up contest. Forty years his elder, Raymond accepted the challenge, then upped the ante, suggesting they do their pull-ups on a door frame instead of a pull-up bar. You could see the bold Coastie's confidence deflate like a three-day-old balloon. The swimmer reluctantly stuck his fingertips on the narrow ledge of the door frame, then proceeded to fall off before getting a single pull-up. Raymond, without saying a disparaging word, gripped the frame and started cranking out pull-ups. He did this until the circle of cheering rescue swimmers were practically bowing before him. It was one of the most impressive physical feats I've ever seen.

After congratulating the new rescue swimmers, the entire tour loaded up onto the C-130 and headed to the opposite corner of the country. It was a six-hour flight from Florida to Washington. I spent most of the time typing up dispatch number two, but I took a quick break to play a round of cornhole. The inside of a C-130 is large enough to hold several humvees, and one of the plane's crew members had set up cornhole boards where the vehicles normally sat.

I spent an hour tossing bean bags with Miss America while traveling 410 miles per hour above the clouds.

Whidbey Island, Washington

The third day of the tour came and went. We made a short stop on Whidbey Island, where we met some of the fighter pilots with VAQ-135. I snapped a few photos of pilots sporting aviator sunglasses and mediocre mustaches, grabbed some quotes about being stationed in the Pacific Northwest, and then loaded back up to fly for Southern California. The short visit allowed me to catch my breath before we moved on to the next stop.

San Diego, California

Once we were wheels down in the Golden State, we were joined by a new batch of B-list celebrities: Wilmer Valderama of *That '70s Show*, Matt Walsh from *VEEP*, and a Christian singer I'd never heard of named Lauren Daigle.

Apparently, Daigle is famous. People worship her, which gave me Jim Jones vibes. She was always smiling, preaching about God's love, and posing for selfies with her crying fans. She's a saint to youth-group types. She was nice, but the way she never stopped posing in the shape of a cross whenever someone took out a camera made me squirm. Daigle, who never stopped talking about Jesus, apparently fancied the way she looked when imitating the death pose of that poor carpenter from Nazareth — arms outstretched in a human T.

Niall and Raymond were both Christian enough that I knew better than to voice concerns over a Daigle-run Jonestown.

Before heading to Texas for the next stop, we visited two squadrons of Marine helicopters. The Marines were quick to show Miss America all around their hulking CH-53 Sea Stallions and V-22 Ospreys. I got the photos I needed, then moved on to aircraft not swarming with Marines trying to catch a glimpse of Miss America in her hot-weather romper.

I climbed into the back of an empty Osprey and sat in the canvas chair closest to the ramp. The aluminum floor plates were clean but stained from a constant state of being wet with oil. It was slick, despite obvious signs the crew had scrubbed the deck in preparation for The General. The nearby trash cans overflowed with paper towels, and the mops still dripped into their yellow buckets. Military helicopters are simply never clean. The first time I rode in one, hydraulic fluid was leaking from the ceiling. Someone noticed my terrified expression and assured me that I only needed to worry if it stopped dripping. "So long as it's wet, you know there's still fluid."

I marveled at the unfinished appearance the insides of helicopters always have. A rat's nest of wires and tubes crisscrossed the ceiling and walls of the flying machine. I looked to my right, up the aircraft's throat to the cockpit, and felt a deep appreciation for the bravery of combat pilots.

Eight years earlier, during my second deployment, I marked a landing zone by swinging an infrared glow stick attached to a string in large circles. The landing zone — which looked perfect on the map — turned out to be a cornfield with 10-foot stalks of unharvested corn. The twin rotors kicked up so much corn dust that I disappeared in the pilot's night vision. If the rotor wash hadn't knocked me over at the last moment, I likely would've been crushed by the 33,000-pound aircraft. Footage taken by a circling F-18 revealed two RPGs soaring past the helicopter as it took off. The white lines left by the rockets were so close that it looked as if they'd passed *through* the helicopter.

Five months after the cornfield incident, I was on another Osprey flight. It was 2013, and the American Consulate in Herat had just been attacked. A suicide bomber drove a van loaded with explosives into the outer gate of the compound. When it detonated, the explosion blew the heavy gate open, collapsed the nearby guard tower with two Salvadoran contractors inside, and shattered every window of the six-story Consulate. The blast was so large it caused a partial collapse of the Consulate's cafeteria roof and left the building's interior in shambles. The bomber's target had been a large generator that powered stadium lights. Had that been destroyed, the seven Taliban fighters who stormed the Consulate in early dawn darkness would not have been gunned down under bright white lights.

After the bomb destroyed the gate and Taliban fighters began firing RPGs into the Consulate, our platoon was called to respond as a quick-reaction force. By the time we were in the air, the surviving Salvadorans had killed the attackers. A team of nearby SEALs also responded. There was no more fighting to do by the time the SEALs arrived, but the presence of bearded, heavily tattooed operators helped calm the diplomats.

As the terrain beneath us changed from arid mountains to Herat's densely populated streets, our flight pattern changed. What had been a calm flight went sideways. Small-arms fire caused the pilot to take evasive action, banking the Osprey at impossible angles. At one point, while I stared out the open rear of the helicopter and did my best impression of someone who wasn't terrified, the ground appeared to spin. It spun until it was above us, where the sun should have been. We were completely inverted. I'd never heard of a helicopter flying upside down, and I was sure we were crashing. The panic that had been rising in my throat with every banking turn suddenly subsided. I *knew* we were all about to die, and momentary sadness was replaced with gratitude. If all of my friends were about to die, I was glad to be with them.

The pilot managed to snake us through the enemy fire, and we touched down on the Consulate's lawn unscathed. We spent the next 24 hours standing guard and picking up body parts from Taliban fighters and Salvadoran contractors. The mission was uneventful, but the flight left a lasting impression on me. It gave me a glimpse into the feeling of helplessness that sustained artillery barrages must bring. The fear of plummeting to Earth in a helicopter was markedly different from anything I'd experienced on my previous deployment. As a grunt, I always had something to do during gunfights and IED attacks. There were more jobs to be done than people to do them, and that kept your mind focused on contributing to the group's survival. When the Osprey started its erratic, fly-like flight pattern, the only thing to do was sit and wait for the ground to slam through the thin aluminum wall.

I climbed out of the Osprey's belly, snapped a few more photos of the country duo, then walked over to a nearby CH-53. I was reminded how much more room was inside a 53 compared to its tilt-rotor cousin. While the flight to Herat and the other near-misses in 2013 stick in my mind like movies, the countless flights aboard 53s blend together. During my first deployment, the 53 was our primary mode of transportation, second only to our feet. They were sturdy, reliable aircraft.

While I sat alone in the silent helicopter, I was hit with an unexpected wave of emotion. All of my memories of being in helicopters had come with a strange sense of surrender. Each flight aboard a screaming, rattling aircraft took a certain level of acceptance. Each flight in Afghanistan was a gamble. Sitting in a helicopter again, this time eerily quiet and still, felt like being inside a giant metal corpse. I exhaled a breath I'd been holding for a decade.

San Antonio, Texas

The second-to-last stop of the tour brought us to the home of Air Force basic training and Pararescue school. The first people we visited were the exceptional airmen who were training to become pararescuemen, or PJs. PJs are among the Air Force's special operations troops and are widely regarded as the most elite search and rescue individuals in the world. They're basically super-medics who can rescue downed pilots from the most remote places on Earth. They're pseudo-surgeons, paramedics who can freefall from 25,000 feet, and warfighters capable of holding their own in a gunfight alongside Navy SEALs and Army Rangers. They're badasses, and Raymond was one of them.

Raymond, now at the very top of the DOD, was an expert in not making anyone feel like their contribution to the mission was any less important than someone else's. A private whose job was to turn wrenches or mop floors would feel like they played a critical role in killing Osama Bin Laden after talking to Raymond. What was amazing was that Raymond believed it. He happened to have a thirst for challenge and adventure, but he never thought of himself as more important than any other service member, despite his maroon beret and current place among the Joint Chiefs.

The only time I saw a noticeable change in Raymond's demeanor was when we went to the pararescue school. We visited the future special operators at their water-survival facility. There, we watched the airmen swim, get screamed at, and swim some more. Raymond took a moment to get some distance from the rest of The General's entourage and talk to each PJ-hopeful individually. He would crouch next to each shivering airman sitting on the pool deck and whisper something in their ear. He looked deadly serious. Afterward, he wouldn't tell me what was said, besides "some words of encouragement." When we left the airmen to continue getting hazed in the water, Raymond's smile returned to his face, where it stayed for the rest of the tour.

After watching future PJs toe the line between drowning and swimming, we moved on to where all the military working dogs in the armed forces get trained. We watched a laser-focused Malinois deliver a bite with 195 pounds of pressure to a young soldier in a padded suit. Then came a tail-wagging black labrador who sniffed out explosives as if he were hunting for peanut butter. He had the oversized paws of a puppy and he stole the show.

It was my first time seeing a Malinois fur-missile go to work up close, but I'd had a lot of time working with those relentlessly happy bomb-hunting labradors.

Two weeks after I got to my unit, I was sent to learn how to become an IED-detection dog handler. The joke in the Marine Corps is that the least reliable Marine in the platoon gets sent to become a dog handler. I hadn't even been there long enough to learn everyone's names when my platoon sergeant told me I was going. He gave me the news by telling me I was selected because of my "uncharacteristic maturity." He spun a yarn about how, while most Marines viewed bomb dogs as little more than squad pets, he'd seen how effective they could be under the right handler when he was in Iraq. Given that I'd only known my platoon sergeant for 14 days, I think the truth was somewhere between the two.

I was devastated to leave my new squad for a month after having just arrived, but I made the most of it. I'd wanted a dog — specifically a black lab — ever since I was in kindergarten and told my parents I planned on marrying my best friend's dog, Sam. The short lifespan of dogs and a few anti-bestiality laws prevented that marriage, but my love for labs never went away. When I arrived at the dog handler's facility, I was paired up with a short little lab named Bonnie. Bonnie had a large dent in her snout from a time she misjudged the distance to the door of her steel kennel and smashed her face. She was low-energy, which meant I had to dance and shout just to get her moving quickly. In spite of her sluggish demeanor, Bonnie never missed an ounce of explosives. She was a rock-solid bomb dog.

Four months after I graduated from the dog handler's course, our unit flew to California for our culminating pre-deployment training in the Mojave Desert. Bonnie flew out to meet us. The first few days went off without a hitch. Bonnie was her usual slow and reliable self. But as soon as we started conducting ranges that involved explosions, she shut down. The whistling sound of artillery simunitions sent her into convulsions. Her PTS from previous deployments was so bad that if a Marine whistled, she would react as if her life were about to end. We completed the month in the desert, but Bonnie was ultimately deemed non-deployable. I never saw that dent-nosed lab again, but a few months after arriving in Afghanistan, we got another dog: Earl.

Earl was also a beautiful black lab, but with the high energy of a puppy. He was three years old and came with a handler: a combat engineer named Red. Red and Earl were a match made in heaven. The two of them were a better Marine-Dog combo than me and Bonnie ever could have been, even if she hadn't been a PTS basket case. In most ways, Earl was perfect. He found IEDs, calmly stuck to Red's side during firefights, and kept everyone's morale up between missions. His only pitfall was that he jumped into wells whenever he was hot. More than once, we had to form a human chain, lying on the ground and holding each other's belts so Red could reach down into Afghan wells and pull Earl out by the scruff. Everyone loved Earl — except for other dogs.

Most homes in Helmand Province had at least one guard dog. Known as Kuchi Dogs, these Afghan mastiffs are unique to Afghanistan and parts of Pakistan. Kuchis are generally organized into two categories: the Tiger-type, which are slightly smaller and walk around with their heads hung low, and the Lion-type. The Lion-type are larger and walk with their bear-sized heads held high, eyes to the front. Lion-type females typically weigh between 85 and 120 pounds. Males frequently grow beyond 170 pounds. Their coat is short, usually some shade of yellow, and has a thick underwool not unlike that of a purebred labrador. A stripe of wiry hair runs down their spine, like on a Rhodesian ridgeback. They have abnormally large teeth for dogs, with fangs that stretch longer than 1.25 inches. Like most dogs, Kuchis' fangs curve into hooks; however, their canine teeth are straight like wolves' teeth. Although internationally recognized as a distinctive breed, Kuchis are considered a "primitive breed" and are therefore banned from competing in dog shows.

Kuchis take their name from the nomadic Khilji Pashtuns, who've bred them for thousands of years to travel alongside them as guard dogs. The Pashtun name for them is Jangi Spai, which means "fighter dog." Pashtuns traditionally cut the Kuchi dogs' ears and tail off while they are puppies. The practice seemed cruel at first, but I came to view it as no different than when Westerners crop the ears and tails of Dobermans and Rottweilers. Their earless, square heads — always high and alert — make them downright intimidating. With their aggressive temperament, Kuchis are a breed you simply do not fuck with. We took to calling them Lycans, after their resemblance to werewolves.

THE DOG AND PONY SHOW

The Lycans barked anytime our patrols came near their fields, night or day. They made the risk of setting night ambushes not worth the reward. More than once, we braved walking through IED-saturated farmland in the dark only to have a Lycan notice us and blow our cover. They were loud, but rarely attacked. As long as we gave them a wide berth and shouted back, they would retreat. Some Marines took to shooting the aggressive dogs, but just throwing a rock in their general direction would usually send them scampering.

On one patrol, in late July 2011, Earl had already jumped into two wells in an attempt to escape the oppressive heat. He was exhausted and overheating like the rest of us. Panting the 110-degree heat was doing little to cool his body temperature, and we found ourselves pausing for a few minutes every time there was shade, just so Earl could cool down. We were on our way back to the outpost when I noticed a tree-lined canal up ahead. A perfectly shady oasis for Earl. He noticed too, and started trotting ahead of the patrol to get there faster. He got about 30 yards in front of me when I spotted the biggest Lycan I'd ever seen. It came sprinting from behind a nearby mud compound and was headed straight for Earl, never making a sound.

Every other Lycan had barked incessantly, but this one was laser-focused on Earl, who was absent-mindedly trotting for the next patch of shade. The Lycan only took two strides before Red and I simultaneously came to the same conclusion: If we didn't shoot this thing before it reached Earl, our bomb-sniffing Lab was going to die. I raised my rifle and tried to track the silent torpedo of white fur cutting through the knee-high vegetation. It had already covered half the distance to Earl. I estimated a lead and started shooting. Red opened up too. The Lycan closed the distance in a flash. The dogs were about to become a yin-yang of black and white fur and I took my finger off the trigger. Red fired one more shot.

The Lycan released Earl, then turned back toward the compound, ran a few feet, and lay down. Its body became obscured by the crops.

"Make sure it's dead, then check its mouth to see if it bit Earl!" Red shouted as he knelt to look over his four-legged friend.

If the Lycan bit Earl, he and Red would have to fly to a large base and get Earl tested for rabies. That would leave us without our two best IED-finding assets for days, possibly weeks. If I found blood and black fur in the Lycan's mouth, I was supposed to cut its head off to send it back for testing.

I approached the animal with my rifle up, half-expecting it to stand on its hind legs and attack me like some horror-movie monster. I got within a few feet and shot it again in the side. Its body lurched, but it remained silent. I set my rifle on the ground and knelt down next to it. With my Kabar, I gently lifted one of its jowls to look for fur or blood. Before I could react, the Lycan reared its head and snapped at my hand. I recoiled, then plunged the seven-inch blade into its neck. It felt like stabbing clay. Only a few inches penetrated its thick coat before the knife hit an immovable wall and jerked awkwardly out of my grip. It was enough to make the Lycan lay its head down on the dirt. I pulled the knife out and stabbed it again, this time in its armpit, in hopes of severing the brachial artery. The full length of the blade slid into the animal's body with ease. Without lifting its head, the dog exhaled as if it were exhausted and just wanted to be left alone. I pulled the knife out, and dark blood trickled onto its pale fur.

I found some of Earl's fur in its mouth, but Red gave Earl a close examination and decided the Lycan's teeth never broke his skin. I was grateful not to have to saw the giant dog's head off, especially after noticing the animal's owners — a farmer and two kids — peeking around the compound wall, had watched the whole thing.

— >><< —

Following the visit to the K9s in San Antonio, we loaded onto buses and drove to where Air Force recruits begin their basic training. The base had arranged a special greeting for the USO tour that simulated the recruits' arrival. Angry female instructors in fancy, Australian-looking bush covers met us at the bus doors. They yelled and stomped their feet, shouting instructions for us to form several single-file lines. Miss America, the band, and the USO staff were giddy for a taste of "real" boot camp. The instructors gave me the impression of people who were doing their best imitation of drill sergeants they'd seen on TV. They shouted with mild enthusiasm for about 15 minutes before finally dropping the charade and asking whether anyone had questions about the rigors of Air Force basic training. As a reporter, I had a lot.

We toured the barracks and were shown the *comfort room*, where recruits could decompress, watch TV, and relax together. When one of the instructors mentioned periodically taking away cell phones, I asked her to repeat herself. She assured me that taking away cell phone privileges incentivized the recruits to work harder. "You want to call home? Make this bed in eight minutes!" she mimicked. It was a far cry from the Marine drill instructors I was used to. My eye caught Niall, whose jaw was also on the ground.

If we hadn't already seen some of the Air Force's toughest members in the pool, I might have judged the entire branch as soft. But the night-and-day contrast between Marine boot camp and the eight weeks of basic training given to airmen each produce a separate desired result. The Air Force fosters a welcoming environment where new airmen emerge eager to pursue a career in the branch best known for cutting-edge technology and good living conditions. In contrast, the Marine Corps guards its traditional recruit training like a starving pitbull does a bone. By design, new Marines emerge aggressive, convinced of their own invincibility, and thirsting for a chance to kill someone.

I took a deep breath, told myself the Air Force knows best when it comes to making airmen, and gathered the quotes I needed for my dispatch. That night, the USO put on its biggest show of the tour.

Killeen, Texas

The tour's last stop was at Fort Hood (now Fort Cavazos): 215,000 sprawling acres on the edge of the hill country in Central Texas. It's home to 40,000 soldiers, including those of the Army's 1st Cavalry Division. Modern cavalrymen ride helicopters and armored vehicles into battle, but that's not what we saw during the tour.

The base commander sent us to visit the cavalry's horse stables. Evidently, the Army pays to have a special detachment of soldiers dedicated to riding horses and performing for crowds. They dress up in uniforms from the 1800s, handmake their own leather saddles, and practice shooting rock salt at balloons with cowboy guns. All of this is funded by the DOD in the name of recruitment. After a day of running around to cavalrymen with my notebook, asking for quotes about why they joined, not a single soldier mentioned horses or popping balloons. In fact, none of them had even known about the special historical unit prior to joining the Army.

Thankfully, the rest of the 1st Cavalry Division was dedicated to perfecting actual warfare. Many of them still wore blue cowboy hats and silver spurs, which they attached to their combat boots. Soldiers who wore spurs walked around as if the rest of the Army were jealous of their Western roots. The reality was that the rest of the Army made fun of the cowboy cosplayers.

The General shook hands with the troops. Miss America posed for photos. The band put on a great final show that doubled as a 4th of July concert, complete with fireworks. Raymond even busted out a custom Air Force electric guitar and joined the band on stage for a rendition of *America the Beautiful*. At the end of the evening, all of the tour members went back to the hotel to unwind over drinks.

I joined them for a beer, but after a week on the road, I was all out of socializing energy. Cozying up to The General and talking about how great the tour was over cocktails felt sacrilegious. Americans were still in Afghanistan; and here I was, throwing back cocktails with some of the highest-ranking officers in the entire DOD. They were nice, but I was out of place. I executed a perfect Irish Goodbye and went back to my room, where I slept like a rock.

The Mountain

> *"The accrued guilt and clutter of day-to-day existence —*
> *the lapses of conscience, the unpaid bills, the bungled*
> *opportunities, the dust under the couch, the festering*
> *familial sores, the inescapable prison of your genes — all of*
> *it is temporarily forgotten, crowded from your thoughts*
> *by an overpowering clarity of purpose, and by the*
> *seriousness of the task at hand."*
>
> *Eiger Dreams,* Jon Krakauer

OUR EXTREME ISOLATION finally set in. It shouldn't have taken me 10,188 feet and eight hours of ascending Mount Rainier through the clouds into blizzard conditions to realize this, but it did. We were on our own, hours away from the nearest piece of dry clothing or warmth, and no one was coming to help. No magic switch could be flipped that would make the wind and rain stop, no portal to transport us back to the base of the mountain, where cars and dry clothes waited. It was up to us to figure something out. If we didn't, we risked freezing to death.

The realization was thrilling. The same part of my brain that missed combat was hoping for this kind of scenario. The rest of my brain — most likely the part not rattled around by IEDs — was getting nervous. Numbness crept from my toes to my ankles, and now climbed up my shins toward my knees. My hands were clumsy from lack of blood, and icy water was running down my back and attempting to breach my waistline. I was painfully cold and getting colder.

The last directions my wife and my editor had separately given to me before I left on this assignment popped into my head: *Don't fucking die up there.* I still didn't plan on failing that directive, but the fact that it even crossed my mind was a bit unsettling. We needed to find shelter, and we needed to do it quickly.

Terrain takes on a new identity when you're walking. If I walk far enough, beyond the first pangs of exhaustion, I can feel the terrain change beneath me. Slight changes in elevation reveal themselves like previously unseen mountains, shooting up through my feet and legs and into my back. Hard-packed dirt becomes downright orgasmic after days of walking through soggy fields. The same is true for loose rocks and sand. Even the slightest instability underfoot can turn an easy walk into a workout. Ask anyone who's ever taken their morning jog on the beach thinking it would be a nice scenic treat, then found their leg muscles screaming at them for the remainder of their vacation.

I can only begin to comprehend the sheer size of the Earth if I walk farther than my body wants me to. I imagine astronauts share the same ant-like sense of divine insignificance. Walking helps me grasp the contradictory magnitude and miniscule nature of our little blue dot.

If I walk for long enough, my brain starts to open up, revealing levels of consciousness otherwise undiscovered. Marathon runners experience the same feeling and call it *runner's high*. Life's stressors fade away as the burning in their lungs and pain in their joints occupies the front of their brain. Eventually, if they keep moving, that acute focus on pain disappears, replaced by a pleasurable out-of-body sensation.

The first time I felt this bizarre walking-meditation was during the culminating event of Marine Corps boot camp, known by a slightly over-dramatic name: The Crucible. Over the course of 54 hours, we marched — or *humped* — 40 miles while carrying 45-pound packs. At the time, 45-pounds felt like an anvil, but our bodies later grew accustomed to humping longer distances with more weight.

During the final stretch of The Crucible, we marched faster than any of us ever had before. By then, we had slept a total of eight hours during the previous 54. Some of the recruits around me marched in a weird state that existed somewhere between sleep and being fully awake. Unlike runner's high or walking-meditation, this state of consciousness is more akin to being undead. Their eyes closed, and they moved forward on momentum alone. My feet and legs screamed for me to stop. The recruit next to me had developed blisters the day before, and every time his foot hit the ground, blood squirted from the two eyelets in his jungle boots. He never complained. None of us did. The end of boot camp and the official transformation from recruit to Marine waited for us at the end of the hump. I'd been drinking the Marine Corps Kool-Aid for three months by then, and I'd have dropped dead before quitting.

Every Marine must complete the 40-mile hump of The Crucible, but for infantry Marines, that's just the foundation for an existence built around humping. The speed at which infantrymen can move on foot has always played a direct role in their success in combat. It was true for Greek Hoplites and it was true for Stonewall Jackson's notorious "foot cavalry." It's still true today. Being able to walk farther and carry more weight than everyone else separates grunts from other Marines and soldiers. That, combined with the knowledge that the infantryman's job culminates at bayonet range in a pair of worn-out boots, is why grunts walk around base like cocks among hens. Through humping, new grunts begin to realize they can endure more pain than everyone else. It taught me how to smile through suffering.

After boot camp, I went to the School of Infantry, where humping took on a whole new meaning. What I'd done at Parris Island was a leisurely walk in the woods compared to the humps at SOI. Every hump was now done in body armor and with the added weight of crew-served weapons. While it still didn't match the difficulty of extended combat patrols, we moved farther and faster than I'd previously thought I could. And as the training progressed, so did my appreciation for the mental catharsis walking provides.

Nine months after graduating from SOI, I was in Afghanistan, putting that simplest of combat skills to the test. We patrolled every single day. The mindfuck of knowing everywhere we went was a minefield combined with the strain of humping a full combat load in the Helmand heat led me to mentally detach. I could pay attention to everything I needed to while also pretending I wasn't really there.

Later in life, I tapped into that ability to detach again.

— >><< —

Twenty-four hours before getting stuck in a snowstorm 10,000 feet above sea level for the sake of a magazine article, I'm twisting my brain figuring out how to get everything into my climbing pack.

Strewn across the lawn of a modest rancher in Puyallup, Washington, is every kind of climbing gear one might possibly need to climb "The Mountain," which is what Seattle residents often call Rainier. Tents, packs, ice axes, crampons, glacier glasses, goggles, freeze-dried food, water bottles, insulation pads, sleeping bags, ropes, harnesses, and cold-weather clothing all lie in the grass like the aftermath of a miniature tornado.

Our team of climbers — several experienced, most inexperienced — come from two organizations: Patrol Base Abbate and Veterans Adventure Group (VAG). Leading the climb are two Army veterans, Justin Matejcek — the founder of VAG — and Austin McCall, a former Ranger and one of the more experienced climbers on the team. The rest of us are mostly Marines. There's Michael Spivvy, a former Combat Engineer who left one of his arms in Sangin; Kevin Fallon, an active-duty Marine and combat veteran; and Jordan Laird, a former Scout Sniper who served alongside Spivvy in Sangin and endured some of the heaviest fighting in the entire Afghanistan War.

"We've still got team gear that needs to get packed up," Matejcek says to no one in particular. He's reminding us this is not a guided climb, and everyone has to chip in to lug the essentials 10,188 feet up to Camp Muir. No Himalayan sherpas or Samwise Gamgees will be carrying the gear for us. Everyone is their own mule.

Matejcek spends most days between May and September exploring Rainier and the other peaks of the Cascade Range. A majority of the time, he's leading other veterans involved in VAG. While Matejcek has only been doing this since 2016, he's no stranger to leading soldiers through backcountry wilderness. Before leading a life as a mountaineer, Matejcek was a sniper in the Army's 101st Airborne pathfinder company. Toting a heavy rifle across the Hindu Kush prepared him for the rigors of leading men up slopes in the Pacific Northwest. He's already taken 128 veterans up

Rainier, with nearly twice as many reaching the summit than the park's average.

He's quick to explain that VAG stands alone in openly not being therapeutic. In a few short hours, Matejcek lives up to his promise of placing us all in an extreme environment where "veterans thrive and most people merely try to survive," as he puts it. But the line between thriving and surviving becomes blurry at high altitude.

Mount Rainier dominates the Seattle skyline. At 14,411 feet, it's the tallest mountain in Washington and the most glaciated mountain in the contiguous United States. Because of its sheer size and accessibility, Rainier has long been a popular destination for mountain climbers. According to the National Park Service, the mountain sees more than 10,000 climbers annually. Of those, about half successfully reach the summit, and most of them are guided.

I'm stuffing an extra collapsible snow shovel into the outer pouch of my 60-liter pack when McCall walks over.

"Everything good, man? How you doin' on weight?" He picks up my pack to give it a test. "Pretty solid. Just be sure to speak up if you need someone to take some team gear off you during the ascent."

Even with the large tent tied to the bottom of my pack, it's still on the lighter end of the spectrum. It's been nearly a decade since I last humped a pack, but the oppressive weight is comforting in its familiarity.

McCall now wears a floral-patterned bandana on his head, but it wasn't long ago that he was sporting a tan beret in its place. He was stationed just down the road from Seattle with 2nd Ranger Battalion where he deployed multiple times to Afghanistan and Iraq as a member of the Army's premier direct-action unit. McCall is tall and skinny, and now has long hair and a beard. To the untrained observer, he looks more likely to whip out a hacky sack than a weapon, but a closer look reveals a scar on his cheek left from an enemy grenade: a souvenir from his time overseas.

I give a few of the other packs a feel-test, and McCall's is significantly heavier than the rest: the first sign of his lingering Ranger tendencies. His gentle disposition stands in stark contrast to his time in 2nd Ranger Battalion, but it's not until later that I learn of his heroics in Iraq. He was awarded a Bronze Star with Valor Device for his actions the day he picked up that gnarly scar. He's humble, and you'd never know about his background unless you pressed him.

Most of the team is here through the help of PB Abbate. PB Abbate was founded in 2020 by Tom Schueman: a Marine

Infantry Officer and former English professor at the United States Naval Academy. The organization's namesake is one of Schueman's Marines, Matt Abbate: a Scout Sniper who was killed in action while fighting in Sangin, Afghanistan. Among the climbers gearing up around me are three of Abbate's former comrades, including Jordan Laird.

"Hey, brotha! I told you you'd be getting a hug when I met you!" Laird says before he wraps me up in an uninvited bear-hug. "I'm Jordan, your new favorite hippie Scout Sniper."

From the moment he hugs me to when he drops me off at the airport four days later, the smile never leaves his face. His positive attitude rubs off on the whole team.

Despite the outward positivity, Laird's been struggling. Following his years in the Marines, Laird dabbled in security contracting but ultimately decided to hang up his M40 sniper rifle. He tells me he's freelance writing for a few publications but is still looking for a more fulfilling career. When another veteran — who is scheduled to go on a different climb after us — finds out Laird was a Scout Sniper in 3/5, he practically berates him with loose connections to Abbate.

"I worked with a guy who went to school with Abbate! Real stand-up Marine, I hear," the guy says to Laird, ignorant of Laird's close relationship with Abbate.

"Nice man," Laird says politely. He nods along respectfully until the guy gets tired and buzzes away to bother someone else.

"That kind of stuff really frustrates me, dude." Laird says to me. "Why do guys feel like they have to validate their service by making up some connection to guys like Abbate?"

I don't have an answer for him, but I nod in agreement. When Rory killed himself a year earlier, there was a tidal wave of false sympathy on the internet. Droves of people commented on various social media platforms, gushing about how great of a man he was and how deep their personal connection had been. It was all bullshit. The massive social media presence he had created ultimately failed to fill whatever void he'd unnecessarily burdened himself with. The forced connections people felt the need to posthumously brag about irked me the same way this guy bothered Laird.

Most of the people on our climbing team aren't aware that Laird not only called Abbate his close friend and teammate, but he was also the one who held the compression bandage on Abbate's severed jugular vein until the helicopter crew took over and forced him to let go. Laird argued with the crew to replace his hand with theirs, but he eventually submitted to their commands to let go and back away from the aircraft. Abbate died shortly after the helicopter took off.

After final checks and inspections for tomorrow's climb, we retire to our respective air mattresses and couches for an early bedtime. The sun is still high in the sky at 9 p.m., but we've got an early start to a long couple of days in the morning. I scarf down another helping of spaghetti for the extra carbs, then pull my sweatshirt over my head for some darkness.

I wake up at 4:30 a.m. to McCall stuffing emergency radios and equipment into his bag.

"Here," he tosses me an emergency avalanche beacon. "Find room for this."

I'm immediately transported back to getting tossed an extra radio battery or smoke grenade before patrol. The familiar pack-stuffing is comforting, like returning to your mom's house and admiring all the unchanged pictures on the wall.

The whole team is up now, moving slowly in scattered corners of the house, making final decisions on what gear and clothing can be left behind to make room for extra snacks. I decide I don't really need my midweight layer, since I still have a rain jacket and a 700 fill down parka. I replace the layer with some extra food and the emergency beacon before grabbing my overstuffed pack and climbing into one of the vehicles.

Rainier has five regularly used routes to the summit. Of the five, Liberty Ridge is the most dangerous. Despite seeing only about two-percent of climbers who attempt the summit, the less-traveled route accounts for roughly a quarter of annual fatalities. We opt for a route on the other end of the safety spectrum: Disappointment Cleaver.

Disappointment Cleaver is an andesite ridge that bisects two glaciers. The route is thought to derive its name from climbers making it to the top of the cleaver, only to realize they're too exhausted to make it to the summit. In order to make it to the top of the cleaver, climbers have to pass through the Ice Box and the Bowling Alley: two areas with dangerous seracs and frequent rockfalls. Despite its ominous name, Disappointment Cleaver remains the most popular route because it requires less technical climbing. Because of this, the route sometimes experiences bottlenecking during peak season. We anticipate avoiding the crowds, since we're climbing during the front end of the summer rush.

As McCall's 2001 Ford Expedition weaves through the old-growth forests that surround the mountain, it's easy to understand why the indigenous people of Cascadia were some of the first sedentary cultures to remain hunter-gatherers in North America. The region contains the only temperate rainforests in North America. There are more shades of green than I previously thought existed in the natural world. The large trees and rich undergrowth grow so thick, I nearly forget we are at the base of the largest mountain in Washington. A gap in the trees quickly reminds all of us that the Western Hemlocks towering more than 200 feet above us are matchsticks in the eyes of the mountain.

As we pull into the upper parking lot at Paradise, the clouds completely shroud Rainier from view. But the mountain is so immense; its presence still dominates the air. You can *feel* its size looming just out of sight.

We're all friendly toward one another. The beginning of a few inside jokes have sprouted, and a shared excitement is starting to bring us together. I rub sunscreen on my nose as McCall eats another spam and rice snack left over from last night.

"No such thing as too many calories going into a climb," he says with rice stuck in his toothy smile. "I'm not even hungry, dude."

I'm not the only one who laughs, then breaks out another last-minute snack.

We step off around 8:00 in the morning, moving from the parking lot to snow in a single step. The climb starts in the first few yards. Each step is equal parts forward and upward. We walk for a few minutes through scattered pines on an easily identifiable path, then stop to make final pack adjustments or drop unwanted layers. Another routine from my days in the infantry I'd forgotten. Walking in the snow isn't quite as frustrating as sand or mud, but it's noticeably different from hard ground. I ignore the self-doubting voice in my head telling me I didn't spend enough hours shuffling up and down Oregon Ridge.

Oregon Ridge is a 1,000-acre park just north of my home in Baltimore. It's mostly woods surrounding a pair of defunct rock quarries. There's a humble nature center, a playground, and a small stage for occasional outdoor concerts. But the most prominent feature of the park is the abandoned ski slope.

In the 1960s, a wealthy insurance executive had the bright idea to clear two tracts of forest for recreational skiing. The slopes were relatively small — only about 140 feet — and the warm Maryland climate meant fake snow had to be sprayed on the hill in order for it to function all winter. After a few seasons, the operation was deemed too costly, and the ski resort closed down. There are still remnants of the old tow rope and lift. More importantly, there are still two swaths of deforested land leading up to the 600-foot ridgeline. Perfect for a Baltimore resident to train his legs for mountain climbing.

I spent three months taking a 60-pound pack to Oregon Ridge and climbing the old slopes. Walking up a steep incline works different muscles than squats, running, or most other lower body exercises. I paired these hikes with an hour on a stationary bike each morning. By the end of month three, my body was in good enough shape to attempt Rainier. So long as the weather held.

On the mountain, I walk third in our line. Matejcek leads the group, and Spivey takes second. He's about a head shorter than me, but because of the steep angle we're climbing, his silhouette stands above me.

This climb is the first time Spivey and Laird have reunited since the teenage engineer lost his arm and was flown off the battlefield a decade earlier. Not only did Spivey recover from his injuries, but he also went on to surpass what most men accomplish with all their limbs. He's an Olympic athlete.

Spivey competed in the 2018 Winter Paralympics in Pyeongchang, South Korea, placing 18th in the banked slalom and snowboardcross. He also placed 11th in the Snowboarding World Championship. Now he's taking a break from snowboarding to climb Rainier.

The rain comes in waves at first, but just as it starts to get heavy, it clears back up. It would be sweater weather if we weren't climbing. A nice, breezy 50 degrees. But the weather devolves as steadily as we ascend. The temperature drops, visibility decreases, and the wind and rain gain intensity with each wave.

The last time I was this miserable walking was not on Oregon Ridge. Only patrolling in Afghanistan's blistering heat topped the awful conditions Rainier threw at us. Those endless patrols through poppy fields were now buried in a decade of memories. Whenever I think about the summer of 2011, significant events like IEDs and firefights force their way onto center stage. But the reality was, the vast majority of that seven-month deployment was spent simply walking around. Thousands of hours of walking have melted into one vague sensation of tiredness. Myriad veterans have written about war's dual nature of being both exhilarating and boring, but the most accurate, one-word summation of war is *exhausting*.

Three hours into the ascent and I'm reminded of that same exhausted feeling I carried on every patrol. It's become clear that the weather on Rainier is not on our side. The water trickling down the back of my legs is an uncomfortable first sign that my waterproof layers have reached their limit and succumbed to the weather. Eventually, water *always* wins. As I accept the fact that the rest of the climb will be done wet, I can't help but smile. I know that the rest of the team is experiencing the same rain-induced moment of misery with me.

The wind blows a combination of snow, rain, and fog in driving sheets that make it hard to see. The snow-covered ground merges with the cloud we're climbing though, and visibility becomes limited to less than 20 yards. I turn to look below me, and a ghostly silhouette shakes its head at me. I can tell by the exaggerated head shake and bouncing shoulders that Laird is laughing. I laugh too, then turn and pay attention to the placement of my next step.

There's something relaxing about climbing. Being exposed to the extreme spectrum of weather in such a remote place makes life's daily problems seem trivial. The mountain has a way of pulling back the bullshit and revealing the superficiality of so many self-imposed problems. It simplifies life in a similar way that war did, and just like in the war most of this group fought in, carefully placing one boot in front of the other becomes the only thing on our minds.

Alongside Matejcek and McCall, a third person in the group has some prior experience on Rainier. John Whitman, also an Army veteran, is quiet and mostly keeps to himself, except when he can't find his tobacco.

"Where are my papers?" he asked a handful of times the night before. Whitman liked to roll his own cigarettes, perhaps to save money or perhaps to give his hands something to do.

Several hours into the climb — as the weather starts to win — the group stops next to an outcropping of rocks. They jut out of the white landscape like the black humps of some

Lovecraftian leviathan. They don't provide any shelter from the wind and rain, but just the variance in scenery makes it as good a spot to stop as any. I sip water and look around at everyone doing the same. Everyone except Whitman.

"You good, bro?" McCall asks Whitman, noticing he has his head down and isn't drinking.

"Yeah," Whitman responds without looking up. "The altitude is kicking my ass."

Altitude sickness does not discriminate. Like a mortar, it comes for anyone, at any time. It doesn't care how many peaks you've bagged before or what kind of shape you're in. Altitude sickness, or acute mountain sickness, can be life-threatening. It stems from a lack of oxygen getting to your brain due to thinning air at high elevations. It's common, and usually involves only mild symptoms like headache and nausea, but those can quickly develop into pulmonary or cerebral edema if you're not careful.

In the military, open water is often referred to as the "great equalizer," because no matter what kind of shape you're in or how tough you are, if you aren't prepared to swim, it will make a floundering fool out of you in no time. Life at high altitude is similar.

Whitman insists that he just needs to drink a little more water, but he's fine to continue. With the weather getting worse and visibility declining into white-out conditions, we decide to keep the group together and continue upward to Camp Muir.

Recognizing that Matejcek needs to stay close and monitor Whitman's condition, and with McCall bringing up the rear to ensure no one gets left behind, Fallon — an active-duty Marine grunt — takes it upon himself to walk point. He seems at ease in the storm. His demeanor is exactly the same as it was the night before when we were warm, dry, and sharing a spaghetti dinner. He may as well have been sitting on a beach somewhere.

Fallon's never been on Rainier, but he's no stranger to climbing. He's already climbed half of Colorado's 14'ers and Mount Whitney: the tallest American mountain outside of Alaska. I watch as he takes the lead and disappears into the white nothingness above us. Spivey naturally falls in line behind him and I follow.

The wind turns from strong to hostile. It now takes our full attention and a significant amount of energy to fight the wind. The increased gusts deny us the ability to rest between steps, as we have to lean into the wind just to avoid being blown into the void. The rain pelting my hood is the only thing I can hear above the howling.

I start to question, *why the hell am I 10,000 feet out of my league?* I don't *need* to be up here freezing my ass off, desperately searching for any respite from the 90-miles-per-hour shitstorm the mountain was throwing at us.

High-altitude climbing has a staggering death rate of 1 in 10. Because of that level of danger, Grant Barnes, the former chairman of the American Alpine Club's Publications Committee, once described it as "the most absurd of all recreational activities." He went on to describe dying while climbing as "sacrificing one's life on the same altar of egoism that causes men to join the Marines." Blindly trudging through the rain up Mount Rainier, surrounded by Marines, it's hard to refute the comparison.

But Barnes, like most people, misunderstands Marines and mountaineers. Most of us were not here because of our egos. Quite the opposite. In the military, we had all been part of a group whose safety we valued above our own. Cherishing the group over ourselves produced a kind of love that few can relate to. I hadn't felt it since leaving the military, and I think the rest of the men voluntarily freezing on the mountain's slopes hadn't either.

After eight hours of leaning into the wind and placing one foot in front of the other up the Muir Snowfield, I see a series of dark shapes above me: Camp Muir. A tower of exposed rock provides a partial wall to a guide shelter: a stone hut measuring a meager 10 feet by 24 feet. The camp also houses two stone pit toilets and an emergency shelter, though most of the camp is hidden in the whiteout.

Spivey looks back at me and waves toward the rocks. The wind makes it too loud to shout to one another, but I know he's making sure I see the camp. I wave back and return my attention to my feet. They've been soaked for hours now, and I can only tell they're freezing because my calves feel cold. From my shins down, I feel nothing.

I take a few more steps, then turn and make sure Laird sees our salvation up ahead. He waves, and I turn back to eyeing my feet into Spivey's boot prints. After another 10 minutes of climbing, I stumble into Muir. I find Spivey standing behind the guide shelter, trying in vain to escape the biting wind.

"Why aren't you inside?" I yell, even though we're now standing inches from each other.

"No room! They can't let anyone else inside!" Spivey shouts. "Said it's strictly for the climbers who paid for a guided climb."

I can't believe it. We're soaked and freezing. The wind is going to make setting up a tent nearly impossible, and we have a whole team of climbers who are dangerously close to getting hypothermia.

"Did you tell them how wet we are?"

Spivey nods, then shrugs. I stand next to him trying to think of what to do, but the wind and cold leave me with no good ideas. Laird arrives next, just to be equally crushed by the realization there isn't room for us in the shelter. The three of us shiver together and try to come up with a plan.

"It's hard being hard," Laird offers, his smile covered in a wool gaiter.

McCall arrives next. He's not panicked, but the concern on his face is undeniable. He recognizes the seriousness of our plight.

"Whitman is struggling." He says with some concern. "When he gets here, pile into one of those stone shitters until I figure out what we're gonna do."

It's not a good plan, but it's a plan, and usually, some form of action is better than inaction. It doesn't take long for Whitman to show up, and it's clear by his stumbling gate that he's deteriorating. We take him into the stone toilet, and the four of us squeeze into the plywood box no bigger than a porta-potty. All of us are wet and shivering hard. A few minutes later, McCall sticks his head inside.

"Good. You made it, Whitman. Drink some water. We're gonna have to take shifts going out to get this tent set up, but in the meantime, wait here until the rest of the team shows up. I'm going to go back down and look for them," he says before disappearing back into the night.

The rest of the team starts to arrive, and eventually, we're all crammed inside the same pit toilet. If we weren't so cold, we would be laughing at the absurdity of our shitter-shelter. After a few minutes of shaking in the pit toilet, Laird and I decide to venture out and see whether we can find the last few stragglers. The wind hasn't let up at all, and the sun has gone for the night. We notice a headlamp coming down from the higher portion of Camp Muir. Despite all the snow, I can tell the illuminated face belongs to Fallon.

"Boys! I found an emergency shelter. Follow me," he says before turning around and climbing back into the darkness.

We grab Whitman and Spivey and head in the direction of Fallon's light.

Around a snowdrift is the stone shelter. The small, heavy door with a section of steel pole in place of a doorknob looks like the hatch of an old frigate sunken into the mountain. I hunch over and squeeze through the small opening to find two shivering climbers who passed us a few hours earlier, sheltering inside.

"Glad to see you guys made it," one of them says. "Dude, we would've fucking died if this place wasn't here."

The two climbers inside are young, probably teenagers. Based on the pile of soaked snowboarding clothes on the icy floor, his self-assessment seems accurate.

The inside of the rectangular shelter is freezing, but it's out of the wind and rain. A thick layer of ice coats the floor, but two levels of plywood line one of the walls like giant bunk beds. One by one the rest of the team piles in until we're all accounted for. A few short comments of gratitude get muttered, but for the most part, we each silently go about shedding our wettest layers and climbing into our soggy sleeping bags.

The risk of hypothermia is virtually gone inside the shelter. With no unrelenting wind and rain, I finally have a chance to appreciate the moment. Wrapped in my waterlogged cocoon, I realize the mountain has given me exactly what I wanted out of this trip. I stepped off with my eyes on the summit, but I was here for the journey. I wanted hardship and shared misery. I wanted a reminder that I could still endure what others couldn't. A beautiful climb from base to peak would have provided some gorgeous photo ops and some bragging rights, but I would have left feeling cheated. Just getting to Muir was every bit as challenging as I'd imagined it could be. I lay awake smiling and shaking for hours before warming up enough to doze off.

— >><< —

Just like the morning after a bender in a shared hotel room, one by one people start to open their eyes. Quiet conversations slowly elevate until most of us are awake, but we're all too cold to venture out of our bags.

"Dude. I'm not gonna lie to y'all. This fucking sucks," McCall says, the first to address the group. "This sucks more than Denali did."

McCall has climbed Denali more than once, but he's referring to a climb he made in 2011. Denali sits at 20,320 feet, making it the highest peak in North America. In 2011, McCall and four other Rangers decided to plant a 2/75 flag on top in honor of their comrades who died in Iraq and Afghanistan.

Above 17,000 feet, their team encountered two climbers in desperate need of help. The two injured climbers had fallen nearly 1,000 feet while roped to two more climbers who died during the fall. The Rangers helped stabilize the two injured survivors and coordinate a helicopter rescue. For their actions on Denali, all five Rangers were awarded the Soldier's Medal: an award given for life-risking heroism that doesn't involve combat.

McCall starts melting snow to make food. We've each carried more meals than we plan to eat, but this is the first chance we've had to eat something substantial since the previous morning. I shovel piping hot Cajun chicken and rice into my mouth and feel a wave of energy surge through me.

"Who here still has any dry clothes?" Matejcek asks, sitting up in his orange sleeping bag.

A quick survey reveals the entire team is soaked. With no dry clothes, it's clear that a push to the summit is no longer feasible.

Despite the fact that we embarked on the climb with the goal of standing on top of the summit, there isn't much pushback from the group. We all recognize the reality of the saturated state we're in. We eat, drink coffee, and pack up for the descent. Disappointment Cleaver lives up to its name, but the overall mood is positive. The dry clothes and beer waiting at the bottom of the mountain make descending feel like a victory.

While I pull my frozen boots over my feet, I'm hyper aware that most climbing accidents occur during the descent. It's typically when people are more tired and less focused. On Everest, nearly three quarters of climbing fatalities occur while people are climbing down the mountain. I keep that knowledge in the back of my head, but I'm anxious to get my feet into warm shoes.

Forcing myself to keep my mind on my steps takes twice as much concentration as it used to. The wind is still ripping across the ice, but it's eased up some. The driving rain wanes from cyclic to harassing fire.

We make final checks and prepare to begin the long climb down to Paradise. I turn to take in one last glimpse of the summit from the 10,000-foot viewpoint, but the mountain still refuses to fully reveal herself through the clouds. I turn back and trudge after Spivey down into the cloud layer below us.

— >><< —

The day after our team gets off Rainier without a hitch, another climber is killed on the mountain. Unusually warm weather weakened the snow, and the climber broke through a snowbridge, falling 60 feet to their death. The terrain was too rough to conduct a rescue in the dark, but thirteen volunteers and three park rangers were able to recover the body the following morning.

I call Laird a month later to help me remember the foggy details.

"Man, I feel like I can't remember anything besides being cold, wet, and happy at Muir!" Laird says. "I just know mutual suffering gets me going. Whiteout conditions, 90-mile-per-hour winds, being soaked — I was so stoked on that. It was better than summiting."

I try reaching McCall, but he's already disappeared back off the grid. After our climb, he wasted no time getting back on Rainier. He led another group of veterans up the mountain a few days later. Half of them reached the summit, and all made it back down safely. Spivey also wasted little time before getting back in the snow. He thawed out for a few days, then immediately set about preparing for the Beijing Olympics. After surviving having his arm blown off, he can't sit still longer than 10 minutes without feeling like he's wasting time.

I call Matejcek and probe him for quotes to use in my article. My editor is looking for a dramatic tale of damaged veterans overcoming trauma together, and my leading questions probably reflected that. At one point Matejcek gets frustrated and cuts me off.

"You know, people always try to label our climbs as some sort of therapy. The truth is, sometimes we climb the mountain just because we want to climb the damn mountain."

Going Cigaretting

"War is life multiplied by some number that no one has ever heard of. In some ways twenty minutes of combat is more life than you could scrape together in a lifetime of doing something else. Combat isn't where you might die — though that does happen — it's where you find out whether you get to keep on living."

War, Sebastian Junger

THE WARM AFTERNOON air is starting to cool when Sebastian Junger emerges from the TownPump in Thompson Falls, Montana. With a paper cup of gas station coffee in hand, he takes a spot next to me on the curb, where we wait for the others to finish buying snacks.

We both hope the caffeine will give us a bump for the evening's activities. The vice we really lust for isn't caffeine though. What we want sits behind the counter organized in rows of brightly colored paper boxes. Boxes whose once-iconic logos are now obscured behind warnings of birth defects and throat cancer. The Oscar-winning filmmaker and *New York Times* bestselling author only smokes on special occasions. Luckily for me, this trip meets his criteria.

Out in Big Sky Country, surrounded on all sides by colossal pine-forested mountains, even the self-destructive practice we're fantasizing about feels dreamlike. I haven't filled my lungs with nicotine and tar in nearly a decade, but reminiscing about the wartime habit with one of my literary idols — who also happens to have survived some of the heaviest combat of the Afghanistan War — makes lapsing into the forgotten pre-patrol and post-firefight ritual sound as inviting as a warm hearth in November.

"It's all about moderation," he says. "I don't drink anymore, but I'll smoke the occasional cigarette with friends."

His five-year-old daughter calls it "going cigaretting." Her nickname makes the habit responsible for the annual deaths of nearly half a million Americans seem as innocent as flying a kite. After reminiscing about how wonderful a cigarette feels when smoked at the right time — just as you take your helmet off after a long patrol, for example — we agree to grab a pack to share later.

I quit smoking cigarettes the day I got out of the military, but the chance to share a smoke with Sebastian is all the convincing I need to break the 10-year streak. I go back inside and buy a pack of Marlboro Reds. Sebastian buys loose tobacco and papers "as a backup." His gray stubble reveals his age, though it's apparent the 61-year-old journalist has worked hard to stay in shape. He's a former long-distance runner and high-climber for a tree removal company. Being active is in his blood.

I'm in the small mountain town buying cigarettes with one of my favorite authors thanks to PB Abbate's book club — the same organization that sent me scrambling up Mount Rainier a few years ago. We meet on Zoom every month to discuss a new book and meet once a year in Montana to rest, refit, and nerd-out about words. Sebastian is this year's guest. For budding journalists and writers, he's a celebrity. But he's agreed to come out to Montana after last year's guest, Karl Marlantes, encouraged him to join.

In 2007, Sebastian embedded with a platoon from the Army's 173rd Airborne Brigade in Afghanistan's Korengal Valley. Those 15-months of combat spawned two documentaries, *Restrepo* and *Korengal*. It also formed the basis of his book, *War*. In a less direct way, the deployment led to two more books: *Tribe* and *Freedom*. It's Sebastian's unique experiences under fire that led to him joining our club. It also led to his immediate acceptance into our group.

GOING CIGARETTING

The 20 veterans we are with — now armed with coffees, beef jerky, and candy — pile into a minivan and two Durangos. We pull out of the gas station and head into the hills, where our Patrol Base and a chance to smoke a cigarette wait for us.

It's the third night in Montana before we find the right time to open the pack of Reds. Breaking the seal felt like too big of a moment to rush — the stars had to align before we could rip the plastic off. But the moment finally comes after a long and emotional conversation about war and how to write about it.

"If a book about war doesn't end by talking about the mothers, it's not a real war book. It's just a combat book," Sebastian says from across the campfire.

The comment hits like a bullet to body armor. I think back to my first deployment and how short-sighted I was. I had been ignorant of how the war was affecting people back home.

— >><< —

Our outpost in Trek Nawa had a generator, which, in addition to charging our radio batteries, gave us enough electricity to power a few computers and a phone. I could walk back into the tent after patrol and log onto Facebook to chat with friends and family on the other side of the world. Sometimes I talked to them within an hour of launching grenades at strangers. Listening to the daily challenges of life at home made me bitter. I simply didn't care about my mom's unruly students or my girlfriend's arguments with her mother. I lied to myself that I was showing compassion and empathizing with their daily struggles. But I was a teenager. Everyone else's problems seemed trivial. War magnified that youthful self-importance. After all, every step they took back home didn't carry the risk of being their last thanks to a homemade landmine. Once, when my mom asked me what I missed the most about home, expecting an answer along the lines of beer or food, I said sidewalks. The ability to stroll through life without fear of exploding felt like the only real criteria for happiness.

Thanks to the internet, what was once separated by weeks of waiting for letters was now connected by an instantaneous digital footbridge. We didn't completely abandon snail mail. Boxes of letters, snacks, and, if I was lucky, pictures of my girlfriend, arrived every few weeks. However, access to Facebook allowed for instant communication, blurring the lines between the weird world of war and normalcy. It made maintaining the psychological armor required to thrive in combat a Herculean effort.

The generator would regularly overheat, abruptly ending phone and Facebook conversations. At the outpost, when the tent's power cut off, I'd bitch about the imperfections of our lightspeed line of communication, then simply go have a cigarette, work out, or read a book. I never paid any mind to how that sudden mid-conversation silence impacted whoever was on the other end. They had no way of knowing whether the line went dead because of a hot generator or because a mortar landed in my lap. It might be several days before they heard anything to assuage their fear.

I rarely considered how me being in harm's way took a toll on my family. After Cavalier got blown up, I made a point to call my mom. I didn't like using the phone, but I knew she'd be worried sick once she got the news. I thought she might like to hear my voice. When she picked up the receiver, I should have assured her I was fine and downplayed my proximity to the explosion. Instead, I cried. The sound of her only son crying on the phone from the other side of the world probably scared her, but she didn't let that come through the phone. Instead, she tried to comfort me. What I didn't know then was that when Cavalier made it back to Walter Reed Medical Center, just an hour drive from my mom's house, she was there waiting for him.

My mom met Cavalier a few times before we deployed. She knew that, in all of the confusing organizational units I was a part of — a division, a regiment, a battalion, a platoon, and so on — Cavalier would be next to me through everything. The four-man fireteam was the smallest unit and the only one she had a firm grasp on. When Cavalier was wounded, she correctly assumed that I'd been there too. She understood that it was little more than chance that he was wounded instead of me.

When Cavalier arrived at Walter Reed, he was unrecognizable. Tubes sprouted from his throat like the tentacles of some creature dreamt up by John Carpenter. His lower half was gone. The hospital blanket lay flat against the bed where his form should have been. His arms were ripped and actively bleeding into the mummy-like bandages wrapped around what remained. Seeing him so broken must have gutted my mom, yet she continued to return and sit with him most days. For her, it was the closest thing to taking care of me that she could find. Seeing what IEDs were doing to us up close left a small psychological scar that lingered for years.

GOING CIGARETTING

The moms of Marines in my platoon formed a sort of trauma-bonded friend group. They were always in communication with one another. If any of them heard from their son, they would pass along all the information to the other moms. Their little network helped them form a fuller picture of what was going on in Afghanistan. It helped them endure the war together. On occasion, however, their unfiltered lines of communication made things worse.

Two months after Cavalier was wounded, Matt Richard was killed. My mom was entering a drive-through car wash when the news arrived, though it didn't arrive cleanly. Just as she entered the tunnel of giant soapy brushes, one of the other moms called to tell her.

"I just heard Matt was killed. I am so sorry," she said.

The other mom was talking about Richard, but what she didn't realize was that my real name is also Matthew. My mom immediately assumed the worst — that I was dead. She began sobbing. The choking grip of grief and anxiety grabbed her by the throat as suds and water transformed the cabin of her car into a casket of dread. It took minutes of sobbing and asking questions before they figured out the miscommunication. I don't think my mom ever forgave her. She still won't drive past that car wash.

The war's impact on my family extended beyond my mother. When I came home, I married my girlfriend. She endured that first deployment with the stoicism of a Spartan wife. But knowing the dangers of being a grunt in Helmand made my second deployment a year later nearly unbearable for her. Waiting for bits of news 24 hours a day for 212 days frayed her nerves. The anxiety preyed on her psyche and triggered a life-threatening fight with anorexia. When I came home, I barely recognized her. That first hug felt like wrapping my arms around a skeleton. I didn't see it then, but the war was trying to kill her.

— >><< —

Sebastian is being literal when he says a book about war has to end with the mothers. That's exactly how he ended his aptly-titled book *War*. The final page describes a posthumous Medal of Honor ceremony. As the fanfare in the East Room of the White House dies down and the small crowd disperses, the only thing left behind is a Gold Star mother, alone with her grief.

But Sebastian is not just being literal. He's also speaking more broadly about the difference between combat and war.

Most military memoirs, he points out, are really just combat books. They tell stories of men under fire, the hardships they endure, and the sacrifices they make. But combat is just a fraction of the machine. A real war book must include more. It can't ignore the destroyed families and the wasted youth swept up in war's category five winds.

In Afghanistan, those families were often destroyed at the hands of the Taliban. The simple act of pointing out where an IED was hidden could result in a farmer having his baby thrown in a well. The Taliban regularly used rape and murder as weapons to keep the locals from helping us. And while the Taliban were the only ones targeting civilians, we certainly didn't go out of our way to protect them.

Toward the end of the summer of 2011, my attitude toward the locals had shifted. They'd lied to us for six months about whether IEDs were hidden on their property and whether the Taliban were in the area. They would idly watch from their doorways as we'd amble straight toward freshly buried bombs or ambush kill zones. With just weeks left before going home, I took zero risks on their behalf.

Whenever we got shot at, I launched 40mm grenades at every treeline or compound I thought might offer the Taliban a decent fighting position. I didn't care who else I might hit, so long as it helped us win. Of all the grenades I fired in the general direction of the enemy, only once did a grenade land squarely on an enemy position I could clearly see. A cloud of gas gave away a Taliban machine-gun position, and my grenade silenced the gun before the shooter even knew he'd been spotted.

Firefights had grown more intense by then. The Taliban knew we were due to rotate out soon, and they intended on making a statement. They'd stay and fight as we closed in on their positions, whereas earlier in the year, they'd flee as soon as we maneuvered. Toward the end, they continued to fight even if we had air support. In the final month before we left, the Taliban initiated the closest ambush of the deployment.

Poth was carrying an M240 that day and returned such an overwhelming amount of accurate fire that he stopped the attack in a matter of minutes. He shot one fighter's arm nearly off and torched the wheat field they were in before the survivors scattered. This was a stark contrast to our first firefight in January, when neither Poth nor I even fired our weapons. Back then, we were so concerned with overstepping the rules of engagement and obtaining positive identification that we simply scurried around, hoping to see an enemy combatant pointing a weapon at us before we got hit. The criteria for positive identification didn't take into account the confusing nature of combat. It took months of fighting before I realized that the most common targets would not be human silhouettes like those we'd trained on, but rather the little carbon clouds that sprang up around enemy machine guns. The few Taliban I did see up close were either dead or wounded.

The intensity of combat toward the end of that deployment didn't protect us from continuing to carry out the less-sexy jobs that fall to the infantry. We still stood guard and set up vehicle checkpoints. The year before I enlisted, a Marine in the battalion, Jonathan Yale, received a posthumous Navy Cross after standing his ground against a truck filled with explosives careening toward the checkpoint he was manning. He and another Marine didn't so much as take a step backward. Instead, they leaned into their weapons, firing at the driver until the truck exploded. They were both killed, but their actions saved the 150 Marines and Iraqi policemen behind them. A picture of Yale hung in our company office back in North Carolina, and his actions stood as a reminder that even something as mundane as a vehicle checkpoint needed to be taken seriously.

On one of the last days before returning home, we set up a checkpoint based on intel that a white sedan carrying weapons was going to pass by the outpost. We strung barbed wire across the road and searched every white car that drove by. The hours passed without incident. Then a white Toyota stopped a football field away from us.

Four of us watched through our rifle scopes as the driver and passenger — both men — started slowly approaching the checkpoint. Once they were within earshot, they complied with our directions. They turned off the car, got out, and allowed us to search them. When I noticed a long pile of something wrapped in sheets in the backseat, they became nervous. The passenger, an older man with a white beard, refused to open the back door and reveal what was in the sheets. With a muzzle in his face, he relented and opened the door. I ripped back the sheets, expecting to find a pile of rockets and rifles. Instead, a girl about six years old lay motionless across the backseat. With her eyelids half closed and colorless skin, it was obvious she was dead. She had no visible injuries. With help from a translator, I fumbled through an apology, though I didn't know what for. The old man covered the girl's face with the sheet then climbed back in the passenger seat. I never learned what killed her, and we never found a white car full of weapons. It was just another sign of the waste war leaves in its wake.

— >><< —

Brendan O'Byrne, a 173rd Airborne veteran and major character in *War*, sits across the campfire from me. He adds that a book about war must include the unstoppable nature of the man-made disaster once set in motion. O'Byrne describes it as a pressure cooker. Once all the ingredients are placed inside — the soldiers, the weapons, and the motivations — it's impossible to stop the war from within. Outside forces like national leaders and peace movements are the only things that can stop the meat grinder once the sticky switch is flipped. Perhaps the most meaningful thing we, as war veterans, can do, is try to prevent the lid from getting placed on the next pressure cooker.

— >><< —

Preventing war was the last thing on my mind when I was introduced to Sebastian's work in 2013. I breezed through *The Perfect Storm* while lounging in my air conditioned room on Camp Leatherneck. It was an escape from the war and from the desert. His masterful mixing of compelling narrative and thorough research blew me away. I was hooked on creative nonfiction. On a phone call home I excitedly told my wife how incredible Sebastian Junger's writing was. I wanted to read everything he'd ever penned. When I got home a few months later, a copy of *War* was waiting for me. It sent me into a rage, surprising us both.

"What the hell is this guy going to tell me about the war in Afghanistan that I don't already know? I just got back from there a second time," I spat. "This dude was up in the mountains with the Army. No IEDs, just calling in airstrikes whenever they felt fucking nervous."

My ignorance and misplaced anger bounced off my wife's armor. Three years later — then a college freshman carrying a laptop instead of a rifle — I came back to the book. By the end of the first chapter, I realized how wrong I'd been. Sebastian put to paper everything I'd been feeling about the war and, more importantly, how I felt about the people I'd fought it with. The book transcended military branches and Afghanistan's varying geography. It was about the transformative experience of becoming part of a tribe built to succeed in combat.

GOING CIGARETTING

War also forced me to confront all the misconceptions I'd held onto about the Army and about the nature of combat outside of Helmand Province. The soldiers who fought in the mountains and valleys of places like the Korengal, Waypur, and Watapur were among the toughest and most tested troops of the entire war. The terrain left them vulnerable to complex attacks at all hours of the day. In Helmand, I never had to worry about getting shot while taking a shit or about getting blown up inside my tent. For soldiers in the Korengal, nowhere was safe. The average enemy soldier in the north seemed to be of an exceptionally high caliber. Taliban up there regularly attempted to infiltrate American positions or capture wounded Americans. Brazen actions like that were rare in Helmand and never happened where I was.

I'm fiddling with a rock warmed by the campfire when O'Byrne starts describing what fighting in the mountains was like. He speaks with his hands, explaining how every soldier had to transform themselves into an amateur alpinist just to patrol. Every firefight took place on varying elevations, and more times than not, the Taliban held the high ground. Fighting at hand-grenade range was not uncommon up there. When he finishes his thought the group stays quiet. Only the crackling of moisture fleeing the burning logs breaks the silence. It's a nice deviation from the scenario that usually follows a veteran telling a war story. Here, no one tries to one-up Brendan with their own tale of bravado. Instead, we soak it in. I put myself in his boots and try to comprehend getting dragged into the wilderness by Taliban hands. Before my mind wanders too far, Sebastian breaks the silence.

"You know, ever since I started writing *Tribe*, I've been trying to come up with a clear definition of what that word really means," he says.

The flames illuminating Sebastian's stubbled face add gravitas to his words, transforming him from journalist to sage.

"A few days ago, I was standing in the shower and it hit me. Tribe means what happens to you, happens to me."

It's the perfect definition. It's what I felt inside the helicopter over Herat and it's why Broome stood so close to me whenever I was investigating an IED. It's the reason Cavalier took point without being asked and why Poth chose to stand and shoot people's arms off rather than hide. The only thing that makes prolonged combat bearable is the knowledge that everyone is operating under that mentality. *What happens to you, happens to me.* Brilliant.

— >><< —

Sebastian's definition is closer to getting at the essence of brotherhood than any definition I've come across. However, "what happens to you, happens to me" is more about willingness than anything else. After all, when a soldier gets shot, no matter how tight-knit the tribe is, only the one with the bullet hole experiences the effects of a supersonic projectile slamming into their body. When Cavalier walked past me on his way to the IED that changed his life, he was willing to accept any consequences. He embodied *what happens to you, happens to me*. But when it exploded, he endured the consequences alone.

The blast left him feeling as if the two missing limbs were not gone, but broken and bent backward underneath his body. Between periods of unconsciousness, as his friends fought to keep him alive, Cavalier would ask Doc to straighten his legs out. Over and over, Doc pantomimed straightening out the invisible appendages while blocking Cavalier from looking down and seeing the reality. Of those who were there, only Cavalier knows what broken phantom legs feel like. And only Doc knows what it's like to lie and pretend to fix them.

In a fortuitous meeting, I bumped into Cavalier a few weeks before heading to Montana. I was reporting on a Memorial Day event that he happened to be attending. The last time I'd seen him was a few years earlier. He'd been overweight then, chain-smoking cigarettes and drinking too much. His positive attitude had given way to cynicism. He was hard to be around. It was no surprise Cavalier battled a period of over-drinking and over-eating. Surviving the explosion was just the beginning of his long recovery. He underwent years of painful surgeries to shave down coral-like outcroppings of bone growth that sprouted off his shattered pelvis. Twelve years later, he was finally settled into a new normal. He'd improved so much that he was actually healthier than he'd been the day he got blown up. He quit drinking, quit smoking, cooked his own meals, and returned to the perpetually positive homeostasis he'd existed in when I first met him.

— >><< —

GOING CIGARETTING

The campfire smoke orients itself to me. I close my eyes until it moves on to the next person. Ditching tobacco did wonders for Cavalier, but I'm still hell-bent on smoking with Sebastian. After discussing the definition of tribe, the whole group grabs shovels and empty sandbags and heads into the woods. Everyone who comes out to Montana participates in a "sandbag ceremony." During the ritual, each member fills a sandbag and scribbles the name of someone they served with whose actions left an indelible mark on how they try to live their life.

Broome is the first name to come to mind. He drank himself to death seven months ago, but for some reason, the fact that he didn't die in combat makes me choose Matt Richard. I write his name with a black Sharpie, then hand the marker to the next person. For most of the group, it's the first time they've filled a sandbag since leaving the military. Once everyone has their own bag, we meander past the cabin and head into the woods that surround our campsite like a pine sea.

The act of physically filling a sandbag is satisfying. The sound of the shovel's metal head biting into rocky earth is a once-familiar chorus I've forgotten. I fill the plastic sandbag with a shovelful of Montana, focusing on Richard as I do. I think about how he died and how much time has passed.

The sandbag transforms from a weightless piece of plastic to a heavy, awkward object. I tie it off, the bygone sensation of twine in my dirty fingers underscoring all the life I've lived since sandbags and dead Marines were not so distant. No one speaks. The bags are filled without words. I hoist my allotment of dirt onto my shoulder and wait for the others to finish. When they do, we carry our bags back to the campfire.

One by one, we place the sandbags in a circle, fortifying the campfire like a mortar pit. I set mine down, smooth out the wrinkled plastic that bears *Matt Richard*, and move out of the way for the next. By the fourth bag, someone begins to cry. No one speaks. We all wait patiently while he cries. After a minute, he touches the name on his bag with two fingers and stands. Eventually, sandbags encircle the campfire. We find seats on the rocks and folding chairs around us. After a few minutes of silence, someone speaks up.

"Does anyone want to share who is on their sandbag?"

The silence remains. Heavy like January fog.

"The Marine on my bag was an EOD tech," says one veteran with a thick Texas drawl. He describes how his friend was defusing an IED in Afghanistan when it detonated. He survived the explosion, but when other Marines rushed to his aid, a second IED went off. The Texan managed to say through his tightening throat how it still fills him with pride to know it took two IEDs to kill his friend.

When he's done sharing, the only noises I hear are crackling logs and occasional sniffles from the circle of veterans. I stare at my feet, not wanting to make anyone self conscious.

An older veteran steps into the center of the circle, rebreaking the silence. He's at least 6' 3" tall and obviously an athlete. He recently left the Marine Corps after 25 years of serving in the Recon and Raider communities. His vocabulary is that of a man who spent his entire adult life training for and fighting in war, filled with acronyms and jargon. The name on his sandbag is one most Marines know: Douglas Zembiec. Zembiec earned a Corps-wide reputation for bravery after he jumped on top of a tank to help direct its fire during the middle of the fight for Fallujah. He was killed in action three years after earning the nickname Lion of Fallujah. The older veteran describes how Zembiec took him under his wing and mentored him. He's been trying to live up to Zembiec's example ever since the Lion was killed in Sadr City, Iraq.

One by one, more people summon the courage to share. Tears start to roll down my cheeks. Every person has been carrying this weight around without realizing it. I know if I don't speak up and tell people about Richard, it will haunt me later. I wait for the next bit of silence, then try to clear the lump from my throat.

"The name on my sandbag is Matt Richard. We were in the same platoon."

The lump dissipates as I describe the kind of person he was. He was hard on new guys when they arrived at the unit, but he also went out of his way to take care of them. He once gave me a ride after I wrecked my car. He picked me up on the side of the highway, 400 miles from where we both needed to be the following morning. He never even reprimanded me, just said he was glad I wasn't hurt. I spare everyone the details of his death. He'd been protecting his friends when he was killed, and that was all they needed to know to understand who he was.

I pause for a moment to wipe my cheeks.

"I wish I'd written his parents' names on the bag too," I say.

They are both alive and well, but their actions when we returned from deployment speak just as loudly as their son's did in combat. We came home two months after Matt died. We left Afghanistan, flying through Kyrgyzstan, then

Ireland, before arriving in North Carolina. When we finally pulled onto base in our white school buses, it was late at night. Families were gathered in the darkness, waving signs and cheering us like returning heroes. One by one, we filed off the bus and stepped into the crowd of screaming loved ones.

After finding my girlfriend, I recognized two faces I'd never seen before. Matt's face looked like an even mixture of his mom and dad, and when I saw them, I immediately knew who they were. They had the courage to drive from Louisiana to North Carolina and watch as their son's friends emerged from the bus. I will never know the pain they must have felt as the line of Marines coming from the bus stopped, and their son never appeared.

I walked over to them before thinking about what I was going to say. Once I was standing face to face with them, all I could muster was my name and that I was sorry. Just like I'd selfishly done on the phone with my own mom, I started to cry. Matt's mom and dad wrapped me up in a hug and told me they were glad I was home.

A few more people around the fire share stories. The atmosphere is filled with grief, but the prevailing feeling is one of closure. Finally, the silence between stories grows long enough that it's clear the sandbag ceremony is over.

Whispered conversations sprout between pairs until the quiet din of conversation includes a few laughs.

"Ready for a cigarette?" I ask Sebastian.

"No shit," he says.

We stand, stretch the stiffness out of our limbs, and break out the pack of Marlboros. I pull out the first smoke and hand the pack to him as we walk a few steps away from the campfire. Standing next to Sebastian under a starlit Montana sky, we finally go cigaretting. I light mine, take a deep breath in, and think about Matt's Mom.

Acknowledgements

THIS BOOK WOULD never have made it to print without the help of friends and colleagues. Among them, additional gratitude is due to Marty, Ethan, Addie Jo, Tyler, and Keith whose mentorship gave me the tools and the space to write. Special thanks to Tom, Michael, Adam, and the rest of the world's greatest book club. Most of all, this book would not exist without the patience and support of my best friend, Grace. Lastly, thank you to the Marines and sailors of 3rd platoon whose unconditional loyalty to one another taught me how to be a better man.

PREVIOUSLY PUBLISHED WORKS BY DEAD RECKONING COLLECTIVE:

FACT & MEMORY by: Tyler Carroll & Keith Dow
IN LOVE... &WAR: THE POET WARRIOR ANTHOLOGY VOL. 1
WAR... &AFTER: THE POET WARRIOR ANTHOLOGY VOL. 2
WAR{N}PIECES by: Leo Jenkins
LUCKY JOE by: Brian Kimber, Leo Jenkins, and David Rose
SOBER MAN'S THOUGHTS by: William Bolyard
KARMIC PURGATORY by: Keith Dow
WAR IS A RACKET by: Smedley Butler
THE FIRST MARAUDER by: Luke Ryan
WHERE THEY MEET by: Cokie
POPPIES by: Amy Sexauer
ROCK EATER by: Mason Rodrigue
REVISION OF A MAN by: Matt Smythe
ON ASSIMILATION by: Leo Jenkins
SANGIN, THEN AND NOW by Neville Johnson
A WORD LIKE GOD by Leo Jenkins
PHANTOMS by Ben Fortier
KILLERS IN THEIR YOUTH by Nicholas Efstathiou

DEAD RECKONING COLLECTIVE is a veteran owned and operated publishing company. Our mission encourages literacy as a component of a positive lifestyle. Although DRC only publishes the written work of military veterans, the intention of closing the divide between civilians and veterans is held in the highest regard. By sharing these stories it is our hope that we can help to clarify how veterans should be viewed by the public and how veterans should view themselves.

Visit us at:

deadreckoningco.com

@deadreckoningcollective

@deadreckoningco

@DRCpublishing

www.ingramcontent.com/pod-product-compliance
Lightning Source LLC
Chambersburg PA
CBHW071117160426
43196CB00013B/2597